The Butterfly Book

Photography by Andrew Fusek-Peters,
written by Jane Russ.

GRAFFEG

Dedications

This one is for Mike Williams and all
the team and volunteers at Butterfly
Conservation West Midlands. A.F.P.

For Jo Sharplin, who encouraged
me to write *The Hare Book* and
died very unexpectedly whilst
I was finishing this one, my eighth.
An ultra supportive member of the
Hare Preservation Trust Committee
she will be much missed by her hare
chums. Jo loved all of nature and it
is therefore fitting that this should
be dedicated to her. J.R.

Contents

Male Orange Tip on forget-me-not, herald of spring.

Introduction

Back in summer 2018, two incredible things happened that changed my life forever. I was diagnosed with bowel cancer and faced the prospect of surgery and chemo. At the same time, I became fascinated with my local butterflies. Most important was that the word 'fly' was buried in their name and yet there were few photos that described their aerial adventures. I felt their fragile bravery and despite their short and warm season how they persisted and leaped at life regardless. I suddenly had a goal, to capture Lepidoptera as they flexed their wings and took to the air and a photographic love affair began.

Sadly, our UK butterflies are struggling with man-made habitat destruction and climate change, but there are also many good souls out there using science, volunteer time and habitat restoration to keep our vulnerable populations in good health. Here is a book where my collaborator shares a dollop of science, plenty of myths, stories, legends and breeding facts alongside close-up marvels and flighty fabulousness. I came through my cancer and was then lucky enough to spend the last four years delving deep into the miraculous life of butterflies. We both hope you will enjoy dipping and delving into *The Butterfly Book*.

Andrew Fusek Peters

Andrew Fusek Peters brought butterflies alive for us all in his wonderful large format book *Butterfly Safari* and I was really excited that he was up for a small-format butterfly book to fit into the Nature Series. I usually only edit the books I have not written myself so I was happy to agree to co-author this one with Andrew as he had done the heavy lifting with all the photography. I always enjoy the research for these books and hope this will prove a jumping-off point for your own deep dive into the realm of the amazing butterfly. I feel sure this will be as fascinating to you as the other titles in the series.

Jane Russ

Heath Fritilllary.

Grizzled Skipper
on orchid.

What are Butterflies, Moths and Skippers?

What are Butterflies, Moths and Skippers?

The order Lepidoptera has many tens of thousands of species of butterflies, moths and skippers worldwide and, rather like the classification of owls, it is constantly being revised.

Only Coleoptera, the order containing beetles, is larger. For comparison, at the time of writing there are approximately 18,000 butterfly species but 140,000 moth species. The UK has 59 species of butterflies but over 2,500 moths, of which there are 42 superfamilies. Butterflies in comparison just have one: Papilionoidea.

There are one or two exceptions but as a general rule, here are six quick pointers to help you decide if you are looking at a butterfly, a moth or a skipper.

Butterfly (superfamily Papilionoidea)

- Mainly diurnal (active in the day)
- Have a thin, narrow, long abdomen
- When closed, wings are above body
- Colours are clear and bright
- Antennae are straight with a nob on the end
- Form a hard smooth chrysalis for metamorphosis.

Right: Small Tortoiseshell.

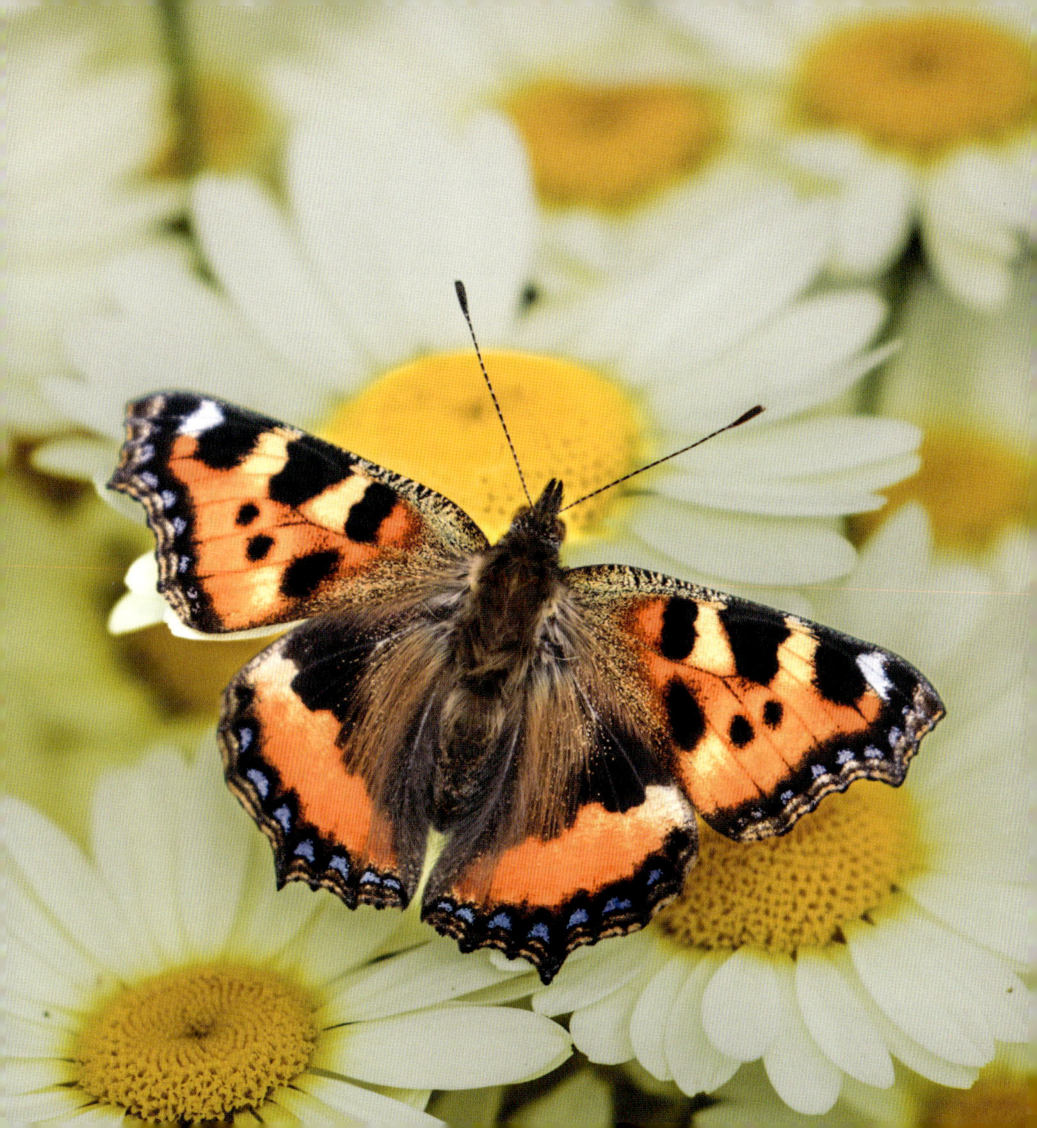

Hummingbird hawkmoth.

Moth (42 superfamilies)

- Mainly nocturnal (active at night)
- Have a thick, solid abdomen
- When closed, wings are at the side
- Colours are more muted and drab, favouring browns
- Antennae are pointed with 'feathers'
- Spin a soft silk cocoon for metamorphosis.

Skipper (family *Hesperiidae*)

Usually considered half way between a butterfly and a moth.

- Are mainly diurnal – active in the day (like a butterfly)
- Have a stocky abdomen and strong wing muscles (like a moth)
- At rest, wings are either upwards like butterflies or out to the side like moths
- Colours are drab favouring browns and greys, with occasional bright ones (like a moth)
- Antennae clubbed (like a butterfly) but are hooked backwards
- Spin a thin silk or silk/leaf cocoon for metamorphosis (more like a moth)
- The main characteristic and the reason for the name skipper is that they have a fast-moving darting and dodging flight. Although more often than not smaller than butterflies, their strong wing muscles make them, at up to 20mph, the high-speed flyers of the order Lepidoptera.

Having set out the differences, there are, of course, always the exceptions, such as moths, that are diurnal and butterflies that roam at dawn and dusk (the term for these is crepuscular), but these lists act as a good starting point to begin to understand the complex and surprising qualities of the butterfly.

Essex Skipper showing
skipping motion in flight.

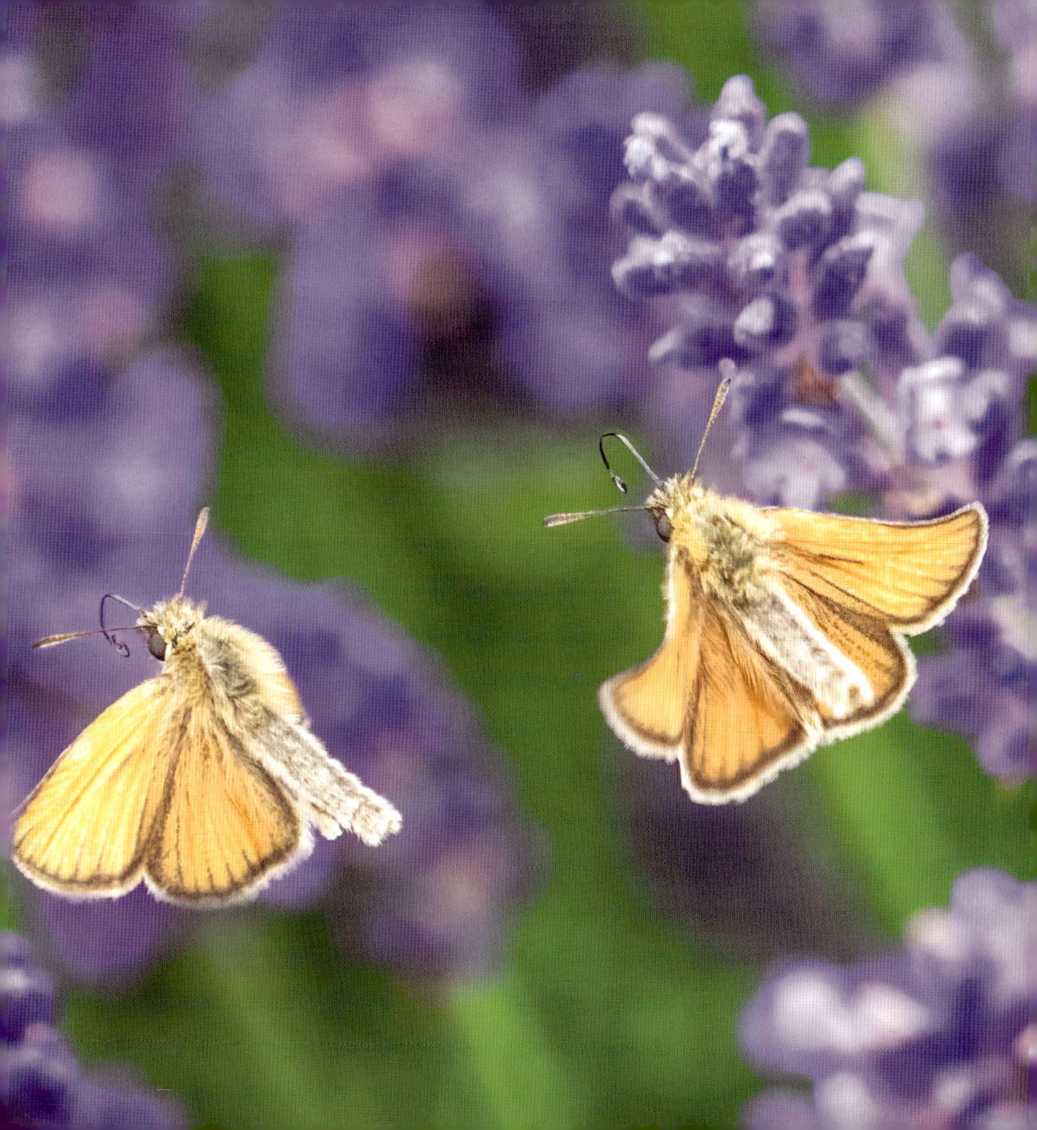

Large Chequered
Skipper showing
proboscis.

Anatomy

Anatomy

The butterfly is no exception to other insects in having three body parts: a head, a thorax and an abdomen. They also have six legs and two antennae.

Butterflies are important in the pollination of crops and garden plants and thankfully, unlike many similar insects, do not bite or cause a nuisance to humans.

The head

The head of a butterfly consists of a pair of antennae, two compound eyes and a proboscis.

- The antennae are the olfactory sensors of the butterfly, picking up from the atmosphere everything from awareness of a possible mate in the area, to flower feeding grounds, to harmful chemicals or other hazards.

- The compound eyes of the butterfly, although proportionally large in their entirety, are made up of many smaller eyes (ommatidia), each having a tiny lens. Information from these lenses is processed by the brain of the butterfly, giving it a huge sweep of vision and awareness of its surroundings and speed, but not in the way that human eyes see a single crisp view of the world. Using specialised photoreceptors, butterflies also have the ability to see ultraviolet light. This is particularly important in detecting some flowers and even

Anatomy of a Butterfly

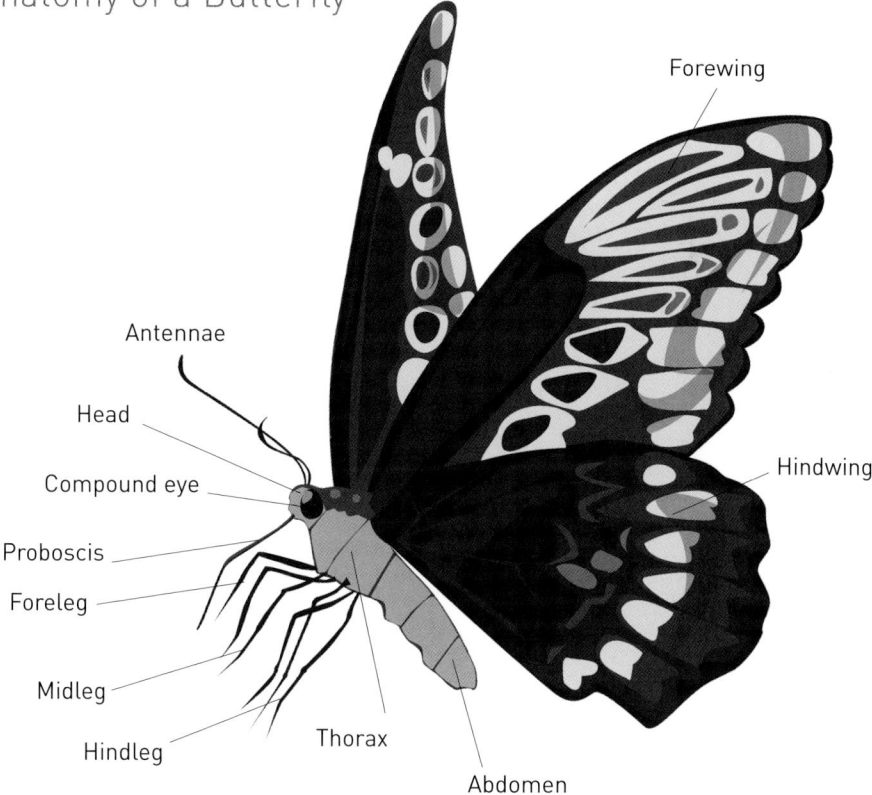

Forewing

Antennae

Head

Compound eye

Proboscis

Foreleg

Midleg

Hindleg

Thorax

Abdomen

Hindwing

Wing scales of Red Admiral.

other butterflies with iridescent elements to their wings.

- The proboscis, which looks like a fine black wire coming out from the narrowest part of the head, has no ability to cut or pierce a surface, it is only used for drawing up liquid nutrients. It can be dipped into flower heads to drink nectar, into tree bark to sample sap, or juices from rotting animals or faeces and, of course, to drink water or fruit juices. When not feeding, it is coiled up like a tiny spring under the head.

The thorax

The thorax is the engine of the insect, it has all the mechanisms for movement.

- Legs. All butterflies have six legs, although in some from the family *Nymphalidae*, the front two are small and vestigial and held close to the thorax. The lower parts of the butterfly leg have names we all recognise: femur (thigh), tibia (shin) and tarsus (foot). The only slight difference from our own is that the tarsus of a butterfly has sensors which give it information about chemicals it walks on. Put simply, they can taste with their feet.

- On first glance, a butterfly wing appears to be just one conjoined wing, however it is actually made up of a forewing and a hindwing. (Moths have a wing coupling device called a frenulum, a tiny bristle that joins the fore- and hindwings together so they beat as one during flight. No European butterfly and most skippers do not have frenula.)

The classification Lepidoptera was first used by Linnaeus in 1746 and comes from the Greek meaning literally 'scaly wing'. The wings are covered with thousands of coloured scales that attach with small hooks like oak shingles on a roof.

These scales can only be seen with a microscope and to the naked eye they seem to be coloured dust;

perhaps you have experienced the colour of the wing brushing off onto your fingers if you have touched a butterfly. There are three types of scales on the wing:

i) Pigmented. These scales have chemical pigments which either absorb or reflect light. These colours will fade eventually as the pigments break down.

ii) Diffractive. These overlapping scales give the wing any brilliant and iridescent colour it might have. The colours do not fade over time and some are highly reflective, like the Queen of Spain Fritillary.

iii) Androconia. These scales contain pheromones used by the butterflies to make contact with other members of the same species.

Butterfly wings are not just for flight: they can be for camouflage, a warning or a distraction to predators and a contact signal to other butterflies.

The abdomen

The abdomen has all the mechanisms for survival.

- The digestive tract. Here the butterfly processes food to support life and the disposes of waste products.

- The spiracles. Whilst they can be found on other areas of the body, most of the spiracles are on the abdomen. These miniscule holes let air enter the abdomen and allow the insect to breathe. There is no facility in its 'mouth' that can be used for breathing!

- The reproductive system. Located towards the tip of the abdomen are both the male and female systems. The butterfly eggs develop here and the female will lay them from the tip when they are ready.

Right: Adonis Blue in flight.

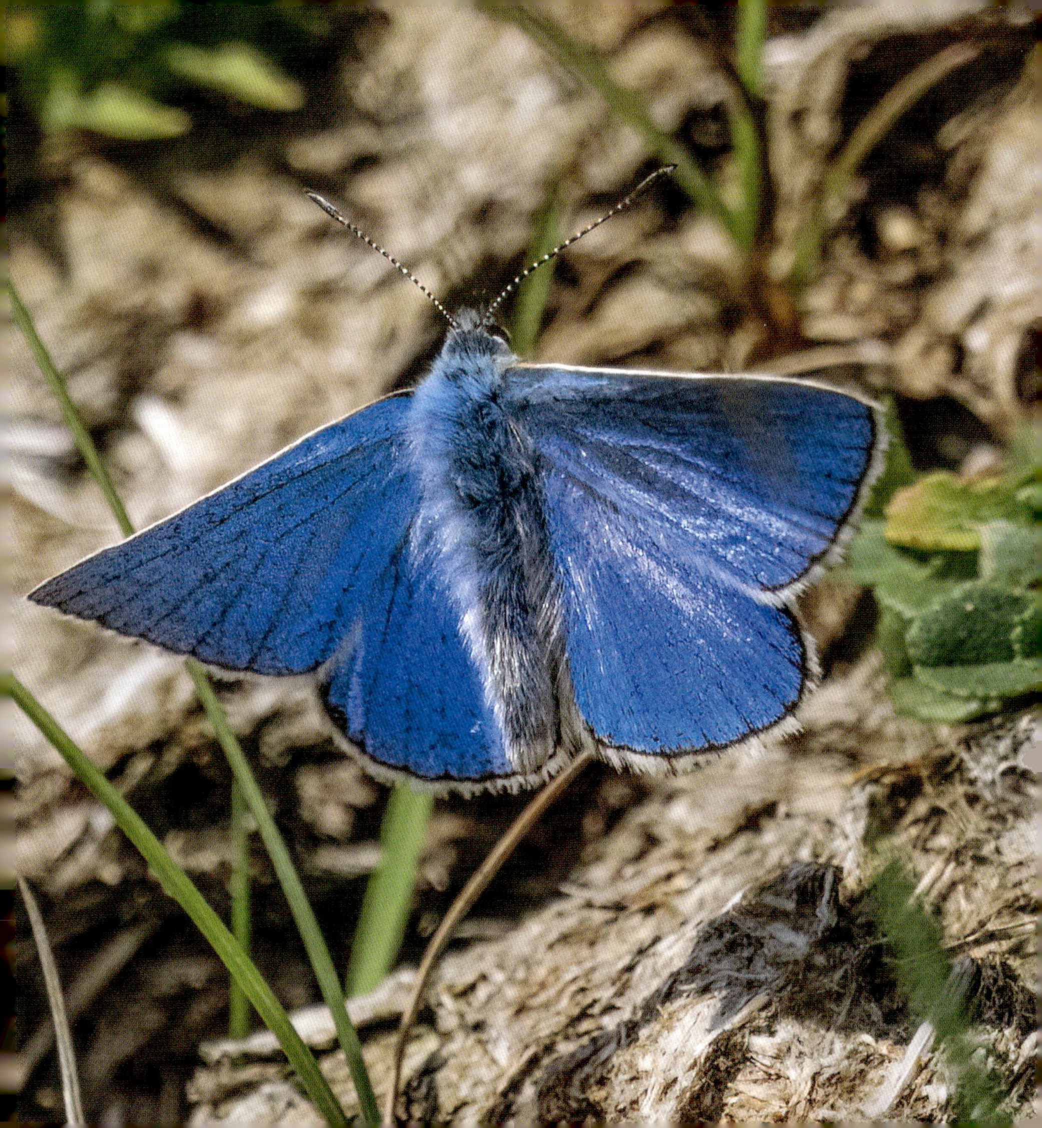

Purple Hairstreak eggs are armoured to keep out predators.

Life Cycle

Life Cycle

Whilst they may only live as adults for a week or two, butterflies have a metamorphic life cycle that has astonished and fascinated us down the centuries.

Butterflies and moths are holometabolous, meaning that they have a complete metamorphosis through four stages (egg, larva, pupa and adult) that bears no resemblance to the final emerging insect. In contrast during hemimetabolism there is a small contrast in each iteration, a good example is the locust, which goes through ever larger nymphal stages, shedding its smaller skin as it goes.

Mating

The reproductive cycle of the butterfly begins with the tail-to-tail mating of the male and female. This is often triggered by the pheromones given off by both participants.

Eggs

The eggs are laid on the underside of leaves or twigs by the female. The chosen site is often a favoured food source plant of the larva, thus ensuring that it is ready to eat and grow as soon as it emerges. It is estimated that only approximately 3% of eggs laid will ever reach maturity. To offset this low success rate, the female butterfly has strategies when laying eggs. A mass of eggs on the underside of a leaf could be a good plan; surely some will survive. However, if that particular leaf falls,

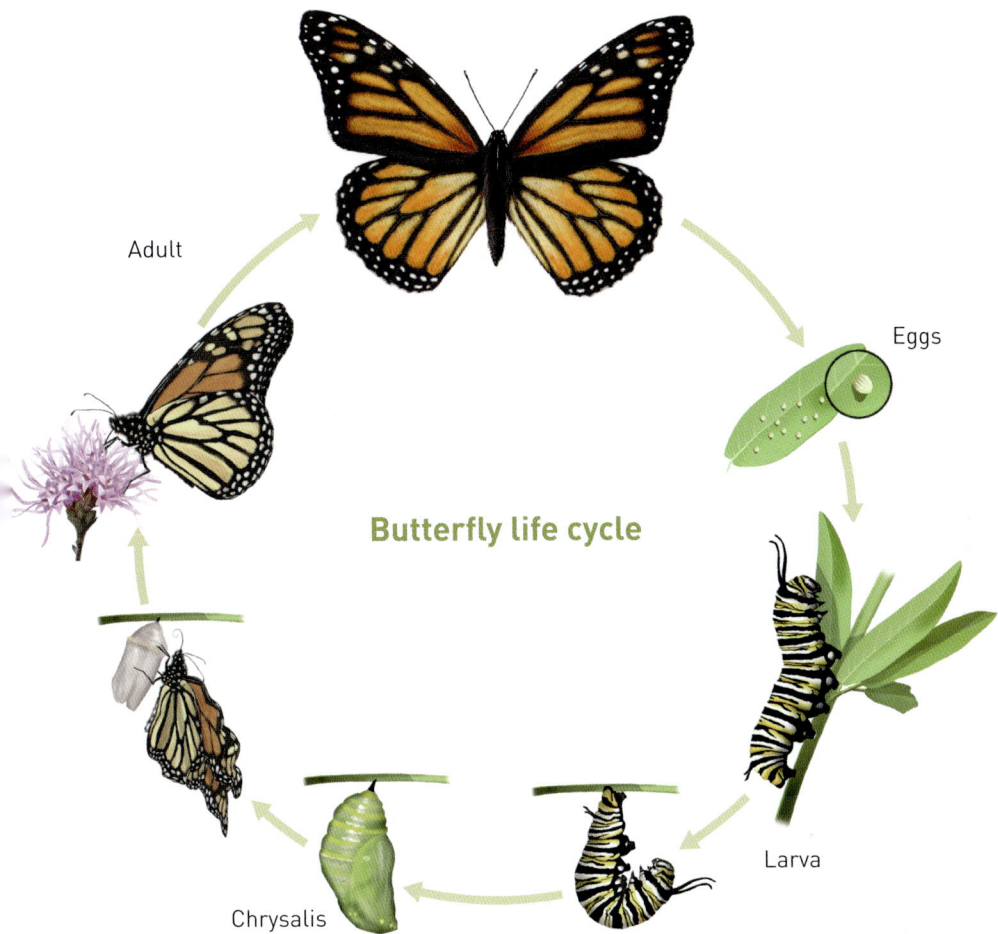

Adult

Eggs

Butterfly life cycle

Larva

Chrysalis

Grayling laying an egg.

then all are lost. Alternatively, laying just one egg in many places spreads the possibility of success and each larva when it emerges has a food source all to itself. Furthermore, a single egg per plant could be very useful in protecting against the cannibalistic instincts of some caterpillars, such as the Orange Tip.

The eggs usually last only a few weeks but if laid at the end of the autumn may, in temperate regions, go through a diapause state or rest, with hatching taking place in the spring when conditions are more appropriate.

The Grayling butterfly in this image (left) has just laid one egg on a twig. The close-up shows the detailed structure, whilst the close up of the Purple Hairstreak butterfly (page 24) reveals its armoured casing, which keeps out predators.

Grayling egg.

Larva/Caterpillar

The larva of a butterfly is essentially an eating machine, chomping its way through its chosen food source, shedding its skin four to six times. The time between each shedding is called an instar and it takes place over two to four weeks, until the larva is large enough to pupate.

The caterpillar has natural devices to guard against predators and these can change between the instars. Spines will put off a bird or small mammal attacks and camouflage means that they cannot be seen in the first place. A nasty smell can also deter those looking for a quick meal as can being a bright colour.

Emerging female
Orange Tip.

Pearl-bordered
Fritillaries mating.

In the natural world an unusual, eye-catching colour can often signal a bad flavour or even poison. If all else fails, there is the strategy of looking like something that is harmful; if you mimic something that tastes bad or smells horrid, that could make you safe too.

Pupa

A pupa has almost no line of defence beyond the siting of them under leaves or in amongst twigs, where 'natural' camouflage will keep them safe. Colours are usually dull and drab to blend in with their surroundings.

The metamorphosis of a Painted Lady

The larval stage of the butterfly is usually known as a caterpillar (above).

This caterpillar is seen hanging under a nettle (right) by a small silk pad and then pupating. The later images show it fully pupated and hanging in the rain. The word chrysalis is used to describe the pupa in this state; it is NOT a cocoon, which is the term used for the spun silk covering of a moth pupa.

The spun silk web of a Painted Lady caterpillar, about to pupate.

The word chrysalis is from the Latin *chrysal* which comes from from Greek *khrusallis*, derived from khrusos 'gold'. Some pupae do have a gold colour or metallic sheen.

The metamorphosis to a butterfly takes place within the chrysalis.

The caterpillar structure breaks down into a thick (often green) liquid (2) and then the cells regroup to form the adult butterfly. In this image you can see the Painted Lady wings inside the chrysalis ready to emerge (3). (The transparency of the chrysalis signals that the butterfly is about to emerge, a useful clue for butterfly photographers that something is about to happen!)

When it breaks free from the pupa it will initially be rather wet and small (4) but, using an abdominal liquid, it will pump up its wings and then wait for them to dry (5). Once dry, they are off (6) and the whole astounding cycle begins again.

1.

4.

3.

5.

6.

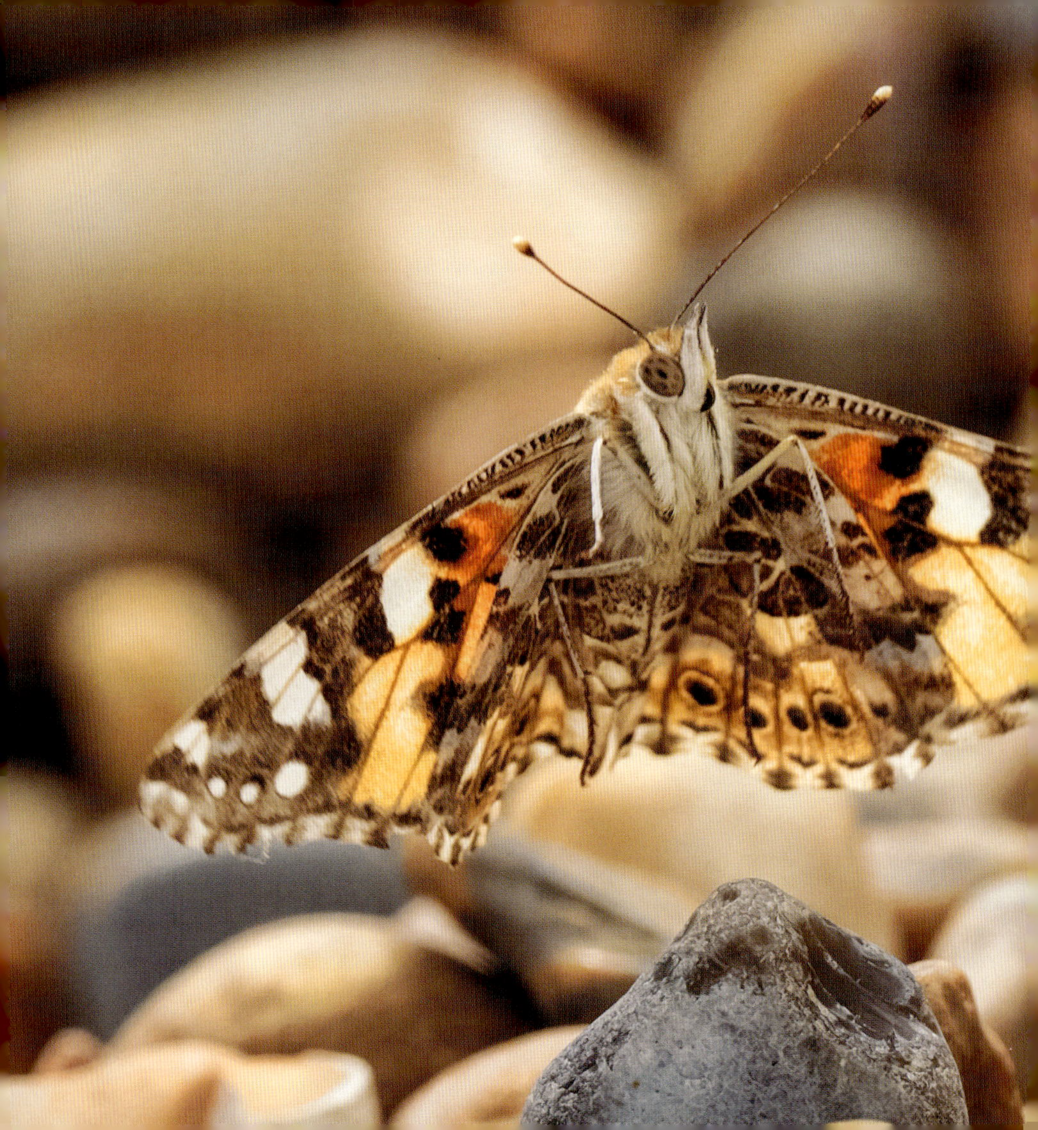

Painted Lady in Shingle Street, Suffolk, just arrived from overseas.

Chequered Skipper.

Five Rare Species

Five Rare Species

All have a Butterfly Conservation priority: High

This book is not about identification of UK butterflies. There are many detailed books about what they are like and where to find them, not least amongst these the amazingly beautiful *Butterfly Safari* by Andrew Fusek Peters (Graffeg). However, it seems sensible to talk about the rarest and also the commonest of these insects found in the UK.

High Brown Fritillary

There has been a well-documented decline of the High Brown Fritillary starting in the 1950s, since when their numbers have dropped by over 90%. One reason could be poor woodland management and the stopping of annual coppicing, which opened up new habitat for this butterfly as old areas became overgrown. Large woodlands in the south, central and north-west of England were its favoured sites and some areas of Wales.

It is now only found in Morecambe Bay, North and South Devon, Exmoor and limited sites in Wales, where conservationists are working hard to keep it from annihilation.

It is a large butterfly with a 60-67mm wingspan. Confusingly, it often flies with the Dark Green Fritillary, from which it is only distinguished by the generally brown colouring of its underwings with, as the name suggests, the Dark Green having mainly green underwings. Brambles, thistles and grass/bracken are principal feeding flowers for both species as adults, with the common dog-violet and hairy violet being food plants for the caterpillar.

High Brown Fritillary.

Heath Fritillary.

Heath Fritillary

It can take the shock of losing something to make people realise they have to make an effort to keep what is left and so it was with the Heath Fritillary.

In 1979 the Large Blue butterfly became extinct and a sharp decline in numbers showed that the Heath Fritillary could go the same way. The butterfly conservationist Martin Warren, following research and having gained full knowledge of the environmental requirements of the species, set about putting a habitat management plan in place that pulled the Heath Fritillary back from the brink.

Although widespread on the continent, it is only found in a few sites in Somerset, Devon and Cornwall and Essex (reintroduced).

This butterfly has a 47-50mm wingspan and, like the High Brown Fritillary, favours woodland coppicing, hence the nickname 'Woodman's Follower'. A low to the ground flitter and glider, it is happiest in sheltered habitats where common cow-wheat can be found, a feeding favourite of both adults and caterpillars. Caterpillars also go for germander speedwell and even foxgloves.

The butterfly conservationist Martin Warren, following research and having gained full knowledge of the environmental requirements of the species, set about putting a habitat management plan in place that pulled the Heath Fritillary back from the brink.

Chequered Skipper.

Chequered Skipper

This small butterfly (wingspan 29-31mm), like others in this chapter, seems to have suffered through poor woodland management, with the replacement of broadleaved trees with evergreens and the decline of coppicing playing their part. It was declared extinct in 1976. However, careful husbandry of habitats by Forestry England and the introduction of specimens collected in Belgium in

2018/19 have meant that Fineshade Wood in Northhampton is now home to a strong and sustainable population. A quick, fast flyer and with markings that make it hard to follow in flight, it is hoped that it will continue to prosper and thrive there. Currently, the Northampton population is the only one in England and there are approximately ten colonies in Scotland too.

The main adult food plant in England is wood small-reed and purple moor-grass in Scotland. The caterpillars prefer both of these too, with False-brome being an addition in England.

Mountain Ringlet

If you see a Mountain Ringlet butterfly, it can be confidently guessed that you are above 350m in either the mountain grassland of the Highlands of Scotland or the English Lake District. The UK's only Alpine species, it is hard to see, with a close-to-the-ground habit. Only flying in bright sunshine, it keeps to

Mountain Ringlet.

Large Heath.

grass tussocks when the weather is overcast or inclement. It is also known to be bog loving.

With a wingspan of 35-38mm, and owing to the inaccessibility and remoteness of its habitat, it is hard to be definitive about food plants. Caterpillars are believed to favour mat-grass and it has also been seen on sheep's-fescue.

Large Heath

The other bog-loving butterfly in the UK, where the draining of agriculture land has been blamed for its decline, the Large Heath has colonies from Wales to Orkney and, with its naturally wetter climate, Ireland too. The declining numbers appear to have been halted by nature reserves with water husbanding their plots, to encourage a rise in the population.

With a wingspan of 41mm, it could be classed as a small to medium butterfly. Interestingly, there are distinct differences in the underwing eyespots in this species, with the northern insects having almost none and the south having very clear spots and intermediate spotting in between. It is possible that this difference is caused by the predation of birds and 'natural selection', although there appears to be no recent research to prove this.

The caterpillars feed mainly on hare's-tail cottongrass, with the adults nectaring on cross-leaved heath. Moorland with a sphagnum moss base is a choice site, as moisture is well held and the food plants can thrive.

The declining numbers appear to have been halted by nature reserves with water husbanding their plots, to encourage a rise in the population.

Small Tortoiseshell.

Five Common Species

Five Common Species

Small Tortoiseshell

A familiar visitor to UK gardens throughout the year, the Small Tortoiseshell is usually the first to be seen in spring and continues right through to the autumn. With a 50-56 mm wingspan, it is in fact quite large and should perhaps be called the Medium Tortoiseshell. Even the name is an anomaly, as when the insect was named in James Petiver's catalogue of 1699, the shell in question did not belong to a tortoise but in fact referred to the turtle. Petiver's 'Lesser Tortoiseshell' and 'Greater Tortoiseshell' subsequently became the Small and the Large.

It shows no habitat preferences and will settle in any garden. The caterpillar feeds on nettles so is not even a threat to gardeners. Whilst the conservation priority is low, there has been a decline since approximately 2020, which is possibly linked to a parasitic fly.

This species is found widely across the UK and even up in the northern part of Scotland.

Red Admiral

Perhaps one of the few butterflies that most people can name in a line up, this is a strikingly coloured and strong butterfly which, with a wingspan of 67-72mm, is categorised as large. Like the Tortoiseshell, it has no habitat preferences and a preferred caterpillar foodstuff is the common nettle. Unlike the Tortoiseshell however, the Red Admiral is migratory, coming initially from North Africa and continental Europe. On arrival in the spring the female lays her first eggs, which hatch in July, the cycle continuing through to as late as

Red Admiral.

Comma.

November. Some adults have been known to overwinter in hibernation. Conservation status is low, with if anything, a slight rise in numbers, possibly due to hibernation.

Comma

Unlike many butterflies in recent years, the Comma (common name Anglewing) has actually come back from a huge decline in numbers in the 20th century and is now widespread in the UK. Its conservation status is now low and it is considered 'not threatened'. Against the trend of other butterflies, its range continues to expand.

The 'comma' of its name refers to the white markings under the wings (right). When folded up like this, the insect could blend into a woody background against a tree trunk when hibernating and, with its frilly edged wings open, it could be a fallen leaf in the hedge mulch. Camouflage is everything in the fragile life of a butterfly. It is thought that this

medium-sized butterfly (55-60mm) declined because its main caterpillar food plant was the hop, which is no longer grown in the quantity that it was in past centuries. Once known as the Hop Cat, the Comma is perhaps a prime example of the flexibility of nature, as the food plant of choice has now become the nettle, which is very prevalent across the country.

Large White and Small White

As the common name Cabbage White suggests, the caterpillars of these two really do love brassicas: cabbage, brussel sprouts and sea kale. This reputation for the decimation of allotments and gardens by the caterpillars belies the fact that they are really beautiful butterflies. The Large is a strong flyer, with successful migrations from the continent. Both sizes are easily seen in most areas of the UK.

The Large measures 63-70mm, with the Small at 48mm, really being more of a medium butterfly. Of course, comparing the size of the two is only useful if you have both in front of you. Here are useful tips for deciding if it is a Large or a Small, apart from size.

The small male and female are both the same, with a slim body form. The colour is a brown/dark grey which is dipped at the point of the upper wing and there is a single or sometimes double dot in the middle on each half. The underwing has a creamy yellowish tinge.

The male and female of the large white however are different, with both having a slightly solider-looking abdomen than the small. The brown/dark grey patterning on the forewings in both sexes runs down towards the head, giving a soft edged semicircular coloured tip to the wings. The female has two spots of the colour on the forewings, whilst the male does not.

Small White.

Holly Blue

The first bright blue butterfly to appear in the spring, flying high above shrubs and trees, the Holly Blue is most often found close to holly in the spring and ivy in the early autumn. Definitely a small butterfly, at 35mm, it is habituated to the south of the UK and Ireland and not generally found in Scotland. Small but elegant, the blue wings are grey edged with shaded dots at the bottom. The wings also have a light fringe of white.

The butterfly shown here is a female, as the male has just a thin black edge to the wings and no dots on the bottom of the hindwings.

Foodstuff for the caterpillar is varied, with holly, ivy and brambles the favourites.

Holly Blue.

Glanville Fritillary
flight sequence.

The Glanville Fritillary

The Glanville Fritillary

An early 18th-century tale of discovery and misogyny.

The queen of natural history in Britain during the late 17th and early 18th centuries, and yet, someone you may never have heard of, is Lady Eleanor Glanville (1654–1709). Like many women naturalists of this time (perhaps with the exception of Mary Anning, the fossil hunter), she had private means and was able to fund her trips and experiments. However, unlike her male counterparts, she was not able to gain full recognition for her work. As a woman, she was unable to even attend a meeting of the Royal Society, let alone become a member.

Fortunately, she was a keen correspondent with several other important naturalists, including John Ray (1627–1705), Joseph Dandridge (1664–1746), James Petiver (1663–1718) and William Vernon (1660–c. 1735). Some of these letters survive and give a clear indication of her ability and the high esteem in which she was held by her male counterparts. Glanville was an entomologist of considerable stature amongst these peers, and James Petiver, to whose *Gazophylacium Naturae Et Artis* (1706) she contributed many first-known specimens, was aware of her importance to his catalogue and gave her a credit in the work. Her submissions went to Petiver pinned, labelled and boxed and were often lesser-known insects, including the first Green Hairstreak butterfly ever documented. On seeing her collection of butterflies in 1703, male naturalists commented it was

Glanville Fritillary. Roosting male and female.

Glanville Fritillary flight sequence.

'the noblest collection of butterflies, all English, which has sham'd us.' A few of her specimens are still held by the Natural History Museum in London.

Eleanor was also a groundbreaker in the rearing of butterflies, obtaining the larvae by beating bushes over a sheet and employing local girls as 'apprentices' to help her do this. She detailed the butterfly life cycle and reared High Brown Fritillaries and Green-Veined Whites and, perhaps most importantly, referenced how she had done it for later naturalists to follow. The techniques she spearheaded are still used today for the collection and rearing of caterpillars but, of course, without the benefit of the temperature and humidity controls we have now. Eleanor was well ahead of her time in this desire to evaluate the progress of the living specimen, rather than just hypothesise about it.

Her greatest legacy and the one that bears her name is the Glanville Fritillary butterfly. On sending her specimen to be documented by James Petiver, he called it the Lincolnshire Fritillary after the area where Eleanor had collected it. However, by 1748, in honour of her outstanding work, it started to become known as the Glanville Fritillary, thus making Eleanor the only British entomologist to have a butterfly named after her. Margaret Fountaine (1962–1940) had a butterfly genus, Fountainea, named after her in 1971, but that is not quite the same thing. The Odonata dragonfly/damselfly species name is *Ischnura fontainei.* The Glanville Fritillary, because of the draining of its original wetland habitat, is now only found on the Isle of Wight and the Channel Islands and very occasionally on the south coast of Hampshire.

It is perhaps what happened after her death that made Eleanor more famous to the general public. To begin at the beginning. Eleanor was wealthy from a young age, as on the death of her mother, she inherited her mother's property and lived alone for ten years. She was eventually married twice; firstly to Edmund Ashfield in 1676, who died after three years, leaving her with two small children and then secondly to Richard Glanville of Lincolnshire in 1685, with whom she had a further four children but, as was typical of the time, only two survived. Richard and Eleanor separated in 1698 (when Eleanor was 44) after Richard had proven to be violent.

Eleanor, being far from stupid, realised that there would be trouble when she died and that Richard might well try to claim some of her wealth. She did what she could to avoid this by putting her property in trust and leaving the bulk of the estate to her second cousin and small legacies to her four children. However, Richard, supported by Eleanor's eldest son Forest, began to fabricate stories about her with the hope of proving her insane.

It all came to a head after her death when, at the Wells Assizes in 1712, Forest attempted to have the will put aside on the grounds of insanity and said that Eleanor believed her children were fairies. Her studying of insects and butterflies was brought up as 'unladylike' behaviour, with witnesses describing how she gathered her samples and how immodestly she and her apprentices were dressed when doing this. Approximately 100 witnesses spoke up and neighbours described her behaviour as eccentric, with her entomological endeavours being called into question. Forest tried to prove that she was insane when she wrote the will and that it should be overturned. It was overturned in 1712, with the judge in the case stating, 'None but those deprived of their senses would go in pursuit of butterflies'.

Footnote: Bearing in mind how badly Eleanor was treated by her Lincolnshire landowner husband

Newly emerged female Glanville Fritillary.

and her discovery in the Lincolnshire Wolds of what was originally called the Lincolnshire Fritillary (subsequently the Glanville Fritillary), the University of Lincoln has an Eleanor Glanville Centre.

This centre is a leading institute working in equality, diversity and inclusion that collects evidence to argue for change in all areas of life and work. I feel sure that Eleanor would approve.

The Purple Emperor

Ask any lepidopterist what the king of butterflies is and without hesitation they will say the Purple Emperor, or 'His Majesty', as the male is known.

This butterfly provides a clear indication that climate change is real. In the past, the season to see it would start in early July and end a month later in early August. Now you would expect to see it from mid-June and it will be over by August.

Living mainly in the high leaf canopy of woods in central-southern England, this glorious butterfly is, at 75-84mm, really large. It feeds on aphid honeydew and tree sap, with only the males coming down to ground level in the mid-morning looking for salts from tracks, moist ground or animal faeces, with dog poo being a particular favourite.

We have noted in previous chapters the decline of a species because of changed habitat. In Victorian times the insect was thought of as common in the south of England, however, the broadleaved forests of the early 20th century gave way to the commercial timber crops created following the First World War. The Purple Emperor did not thrive in softwood glades and so it died out where the broadleaves had been lost. Following the Second World War, the butterfly went into further decline and by 1989 was only found in Hampshire, Surrey and Sussex, with small, scattered colonies elsewhere in the south. Following rewilding, it has returned to the Knepp Estate in Sussex in large numbers.

Its size is not the only thing that marks it out as an 'Emperor'; it has the most iridescent of iridescent wings in a strong, rich purple. This most imperial of colours is not available to see all the time. If you come upon one, it may seem to have black wings with white stripes, however, when the sun catches the wings at a certain angle, they present as the most intense purple-blue you will see in the natural world.

By contrast, the female is a deep brown, with no iridescence to the wings at all.

Not only is it a large butterfly but it is totally fearless, defending its territory by chasing off birds and dragonflies.

There are whole books and websites dedicated to this giant character and for those whose appetite has been whetted, your starting point could be The Purple Empire website at *apaturairis.blogspot.com*.

Butterfly Aberrations

In mammals, birds and insects, the concept of genetic mutations is generally well understood.

For example, red-eyed albinos and normal-eyed leucistic specimens indicate a lack of pigmentation in the skin, fur or feathers. In the butterfly world genetic differences in markings are known as aberrations, a deviation from the normal type, and were first attested to in 1735. One fascinating fact is that aberrations can disappear as speedily as they appear and it raises the question of whether they are caused by genetics or environmental factors. Study is ongoing in this area.

At its most basic level, lepidopterists are always hopeful that what they are seeing is a new, as yet undocumented, species. However, more often than not it is just an aberration of some well-known type. Here are two simple, classic examples to give you an idea of the challenges of even finding an aberration.

The delightfully frilly edged Comma butterfly (left) has a white comma shape on the underside of its hindwings, hence the name. In the Comma ab. *o-album* aberration however, the comma has joined to become a tiny white circle.

Normally, the Gatekeeper butterfly (right) has a brown and orange colouring, whilst the aberration Gatekeeper *ab. subalbida* has lost all the orange hue. Note however, that it has not lost the tiny white spots in the darker brown circles. Furthermore, and confusingly, the Gatekeeper also has an *ab. albidus,* which looks just like the *ab. subalbida* but is more white than yellow.

There are many atypical variations of well-known butterfly types and much information on the internet listing what has been found to date. A good starting place for research is www.britishbutterflyaberrations.co.uk

Painted Lady
on buddleia.

Garden Plants
to Attract Butterflies

Sedum

Garden Plants to Attract Butterflies

There are of course many others, these are just a collection of obvious ones to get you started.

Buddleia

As its common name suggests, the buddleia, or 'butterfly bush', is always guaranteed to draw the insects and many other nectar loving species too. One of life's great survivors, it is often seen on unlikely bits of scrub land, sprouting from the tops of decaying buildings or on railway embankments. If you include it in your garden for the butterflies, be aware it can be a thug unless hard pruned in the spring. Available in pink, purple, white and red, it is an insect magnet.

Sedum

With its ability to withstand a hot, dry position, minimal maintenance and a flowering season into the autumn, sedums are a great entry point for anyone thinking about encouraging insects into the garden. Their almost succulent leaves and many-headed pink flowerets make them a great addition to even the simplest green space.

Lavender

Anyone walking the borders at any National Trust or English Heritage country house in the summer knows the draw of a mass of lavender for bees, butterflies and other insects. Any sunny spot will do and this scented gem can be grown in pots too for those with a terrace garden.

Comma on lavender.

Painted Lady on verbena bonariensis.

Hebe

The strong and resilient hebe is as happy in a pot as in the soil. With a choice of colours from deep blue, purple and white and with leaves in a variety of shades and variegations, it is no surprise that it is very popular with both gardeners and insects. Sometimes rather slow to grow on, it is worth the effort and the butterflies will thank you.

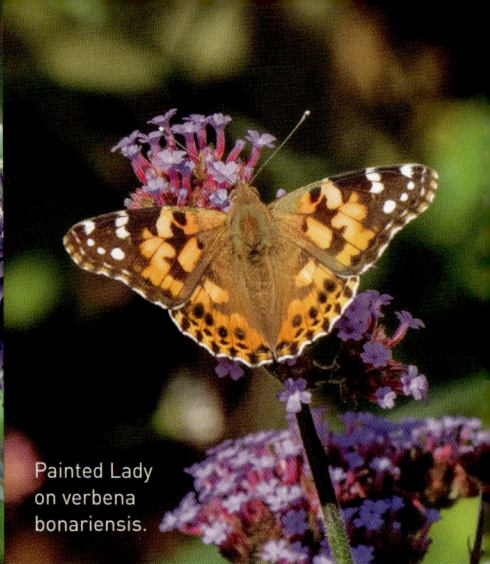

Verbena Bonariensis

A tall perennial for the back of the border, the stiff upright stalks support a cluster of bright mauve flowerets with tiny red centres. Verbena flowers well into October and can therefore provide a useful butterfly food source late in the year.

Echinacea

A choice for the larger garden, the dramatic 7-10cm-wide flower heads are displayed on top of strong, upright stems. Clump forming, they are good as both a cut flower and for butterflies nectaring.

Aster

Another long-flowering plant, offering daisy type vibrant violet/blue flowers with a mustard yellow centre right through from August to the end of the autumn. A hardy perennial, once planted they will offer late food for butterflies and bees year after year.

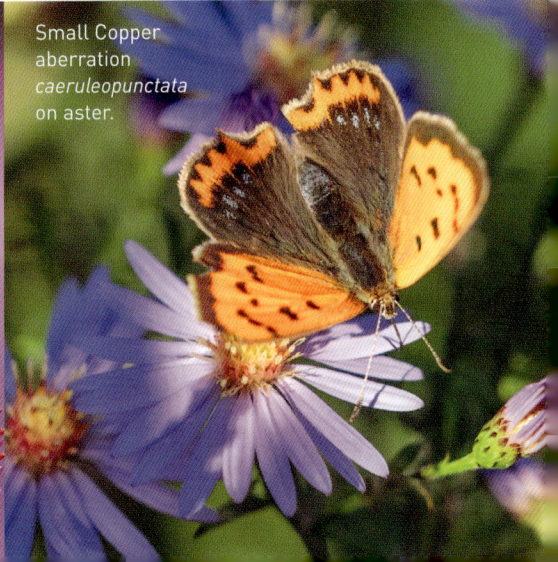

Bumblebee on echinacea.

Small Copper aberration *caeruleopunctata* on aster.

Cornflowers.

Cornflowers

Once a ubiquitous sight in corn fields, these beautiful flowers on their rough, lightly ridged stalks are now favoured in our gardens instead. These days available in white through pink, blue and even dark purple, the strong, bushy plants produce much drama in a border. They are also great as cut flowers.

Fennel

Good for more than cooking, the tall, feisty fennel is a hardy perennial offering late colour in a garden. Another multi-headed flower with a soft yellow colouring, it draws in butterflies and bees just as efficiently as the other plants listed.

Fennel.

Essex Skipper on
lavender flower.

Amazing Facts about Butterflies

Hairy Eyeballs and Other Amazing Facts about Butterflies

The Eye

Butterflies do not have eyelids, their large compound eyeballs containing thousands of ommatidia or separate eyes. These 1,700 individual eyes working as one, give the butterfly very clear vision within a range of about 200m but all are open to the passing airflow as the insect flies and therefore subject to a possible coating of dust, pollen or other matter. To protect against this issue, some insect species have what can only be described as hairy eyeballs. Whilst this has been known about for some considerable time, it was not until 2015 that a proper study was done into the how and the why.

To measure the use and efficiency of hairs, the research team did experiments in a wind tunnel with models built to scale and they also used simulations. The study showed that the length of the ocular hairs is in exact ratio to the spacing of the ommatidia of the eyeball, thus every lens is covered by a hair and as a result the airflow at the surface of the eye was reduced by 90%. The significance of this is obvious: this drop in airflow means that as they fly, butterflies have an inbuilt protection against the depositing of detritus on the eyeball itself, the hairs acting rather in the same way as eyelashes do in mammals.

With so many eyes per eyeball, the butterfly has excellent all-round 360° vision, meaning it has the capacity to feed and watch out for predators at the same time. For comparison, a human has approximately 120° vision.

Right: Eye of newly emerged Painted Lady.

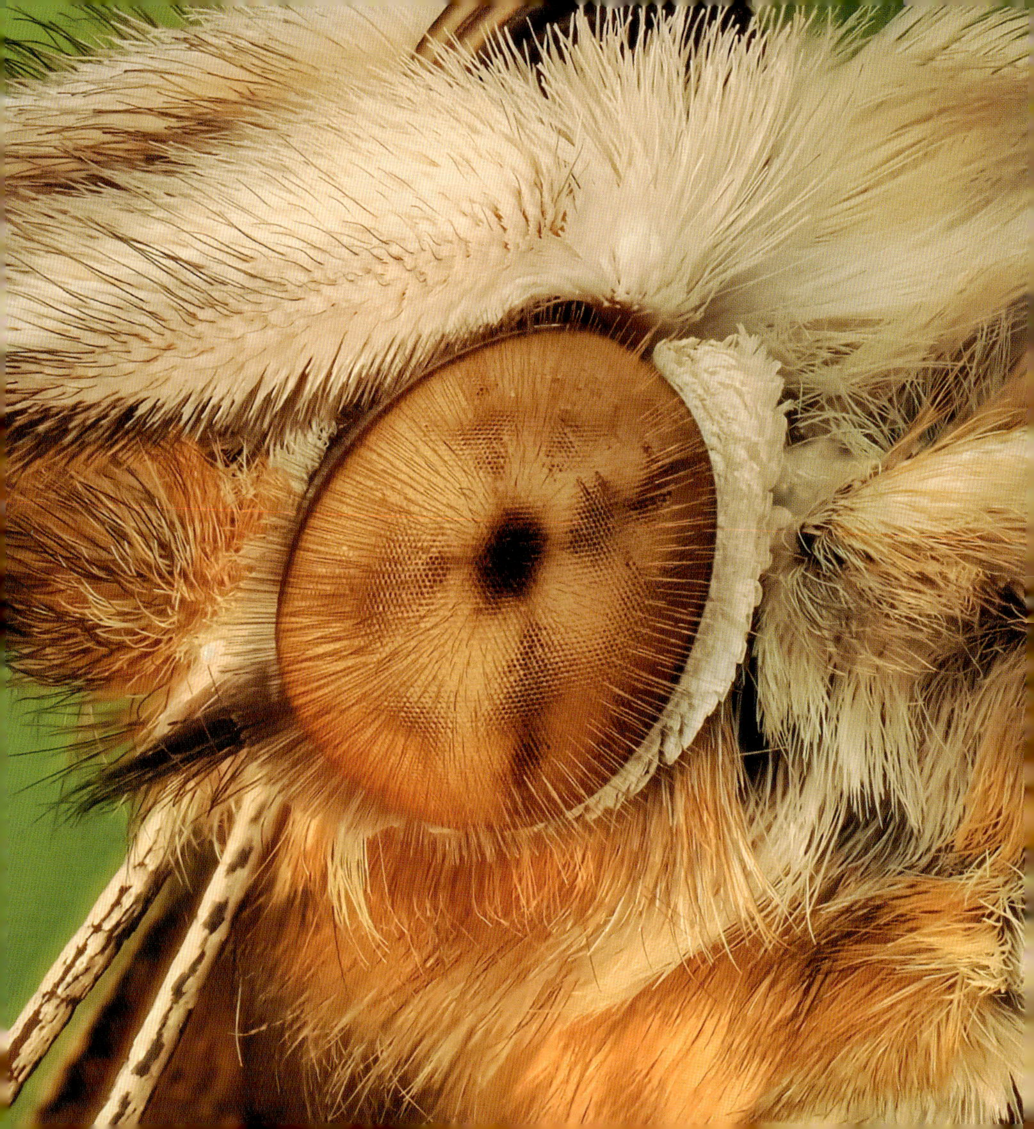

Red Admiral on valerian.

We have an eye with a single lens and rods to catch light and cones, which are photoreceptors, to catch red, green and blue wavelengths. A butterfly however, has in its one big eye 1,700 smaller individual eyes, each with a lens, one rod for light and up to three cones. Our three colour cones are very outclassed by the butterfly, who can see the seven familier rainbow colours and all the intermediate shades as well as detecting infrared and ultraviolet light. Ultraviolet detection is most useful to butterflies, bees and other pollinators as they see the intense colours drawing them into flowers and indicating the quality of the food source.

Movement is also detected well by butterflies – it helps to always be on guard against being eaten. Interestingly, however, some do have issues accurately recognising shapes. There continues to be much research on the eye of the butterfly, with commercial uses, such as keeping outdoor sensors pollen-free to the fore.

The tapetum lucidum is the reflector in mammal eyes that gives them eyeshine at night. Glowing eyes are to be expected when light is shone at the eye when it is dark (it's what Percy Shaw noted in 1934 and caused him to invent the famous cat's eyes road strip). Butterflies have approximately 17,000 lenses on each eye so, do they have eyeshine? The answer is yes, but it only lasts a very short time and can only be seen under laboratory conditions with an ophthalmoscope. Interestingly the pattern of lights reflected in the eye matches different types of butterfly and leads one to wonder if they recognise each other by tapetum lucidum when light levels are dropping.

The Mouth

The mouth or proboscis of a butterfly is like a straw through which it draws up nectar and secretions, as well as mineral salts. The last of these come from muddy puddles or similar 'tasteful' wet areas and are most often frequented by the males. Nutrients gathered this way will be passed on to the female during mating, thus improving the quality and sustainability of their progeny. The butterfly has to assemble its feeding tube itself after it comes out of the chrysalis. It hooks together like two halves of a drainpipe to form a tube. Newly hatched insects are often seen curling and uncurling this amazingly thin pipe to check that all is well. Using the front legs, the proboscis can be removed when necessary for cleaning, thus ensuring there are no blockages.

Butterflies only drink their sustenance and, bizarrely, they 'taste' through receptors on their feet. A female looking for somewhere to lay her eggs will drum on the surface of a leaf to see if the plant juices subsequently released match her requirements. The chemoreceptors on her legs will advise if the chemical combination is what she needs and if so she will lay her eggs on that leaf; if not, she will move on and try elsewhere.

A butterfly of either sex will also step on food sources so that sensors looking for dissolved sugars can record a good source, fermenting fruit for instance.

Mutualistic Symbiosis

Symbiosis is the relationship developed between two species in order to exist. In the form known as mutualistic symbiosis, both parties rely on each other for their continued existence. (There are two other main forms of symbiotic relationship: commensalism – both benefit but neither is aided or damaged – and parasitism – one benefits, the other is harmed.) The perfect example of mutualism is the Silver-studded Blue butterfly and, amazingly, black ants.

During its second instar as a caterpillar, the Silver-studded Blue develops a specialised gland on

Silver-studded Blue just emerged
with attendant ants.

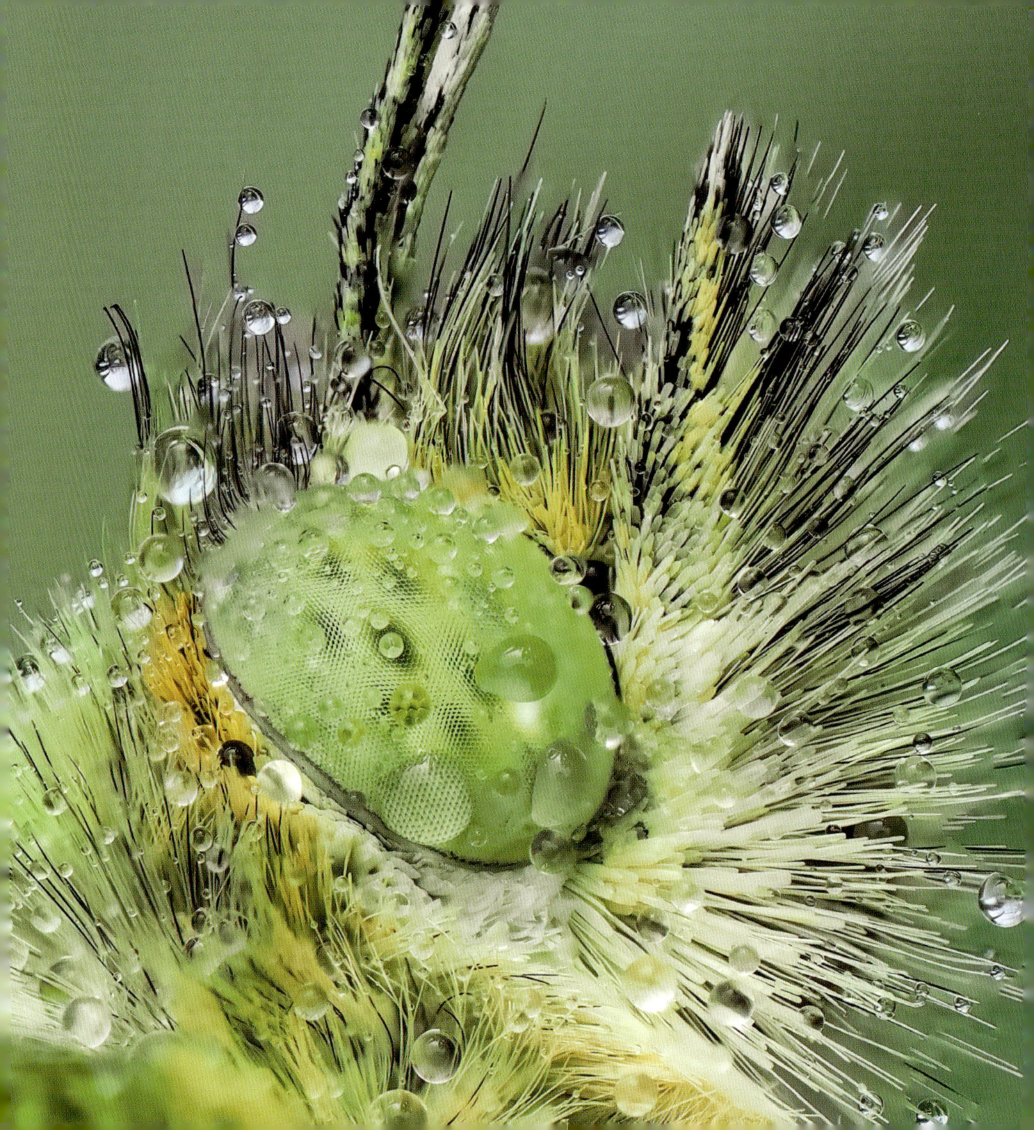

its seventh abdominal segment (newcomer's gland) which secretes a sweet mix of sugars and amino acids. The black ants take the newly hatched caterpillars back to their nest, where they 'milk' them for this secretion. They then take the caterpillars out to feed at dusk. The ants support the caterpillars to pupate and will accompany them during their emergence as butterflies.

Perhaps the most developed example of this close relationship is that of the Large Blue. During a late instar, having fed on plants favoured by ants, the Large Blue uses a chemical mimicry technique to induce the insects to take it into their nest. Once inside, it is either fed by the adult ants or feeds on ant grubs and the pupa is formed actually in the ant's nest.

Left: Small White covered in dew.

Are butterflies waterproof?

Not so much waterproof as water repellant. In 2013, a research project at Boston University found that ribbed textures on a butterfly wing explained why they were so good at staying dry. Water droplets filmed in slow motion and splashed onto a ridged surface flared and splattered and were dispersed 40% faster, leaving almost no liquid behind. This of course, had tremendous commercial potential for fabric, metal and ceramic manufacturers.

This initial research was supported by a further study in 2020, *How a Raindrop Gets Shattered on Biological Surfaces*, proving as before, that micro-scale bumps and a nano-scale layer of wax make the droplets of water shatter, which protects not only a fragile surfaces, like a butterfly wing but also reduces hypothermia risks. A butterfly can only fly when warm so loss of heat in a rainstorm becomes really important.

The lead author of the study, Seungho Kim, explained why they had looked at the implications of high-impact rainfall and said, '[Getting hit by] raindrops is the most dangerous event for this kind of small animal', noting the relative weight of a raindrop hitting a butterfly wing would be analogous to a bowling ball falling from the sky on a human! Slow motion filming of falling droplets showed again that the micro-bumps/ridges on the surface of a butterfly wing act as a needle-type shape, breaking up the water and reducing its ability to damage and cool where it lands. The power to repel water quickly can make the difference to insects and small birds in their ability to fly and thus feed and escape predators.

Above: Photograph taken in the wind tunnel with a hawkmoth, using dyed water to aid understanding.

Chalkhill Blue in the rain.

Peacock and Silver-washed Fritillary.

What Do Butterflies Do in the Winter?

What Do Butterflies Do in the Winter?

Dormancy or Overwintering

In strict scientific terms, butterflies do not hibernate, they are dormant, also known as overwintering. This overwintering can be done as eggs, caterpillars (larva), pupae or adults. By far the smallest number at six, overwinter as adults. Interestingly, they do this using a form of camouflage called crypsis, blending their colours with whatever/wherever they have chosen to rest. These two images show a Peacock at rest in an ice house. I have included an image of what the Peacock looks like in the summer for comparison (page 97).

The butterfly finds a cool, dark spot to rest whilst it is very cold outside, in a tree crevice or a dark shed or workshop. The Brimstone, Comma and Small Tortoiseshell as well as the Peacock, all overwinter as adults.

However, this idea of overwintering has an upside and a downside for the butterfly. The upside: as soon as the temperature rises, you are ready to come out to find nectar and mate. The downside: all the animals and birds who have struggled through the winter are keen for a quick passing food source. Life is, as always, a balancing act.

Of the other UK species that are dormant through the winter months, 9 stay as eggs, 32 as caterpillars and 11 as pupae. One or two even have a choice of which mode to use for overwintering – the Speckled Wood butterfly can be dormant as either a caterpillar or a pupa.

Butterfly Conservation (BC) note that only the Small Tortoiseshell and the Peacock are frequently seen overwintering in our homes.

A dormant Peacock.

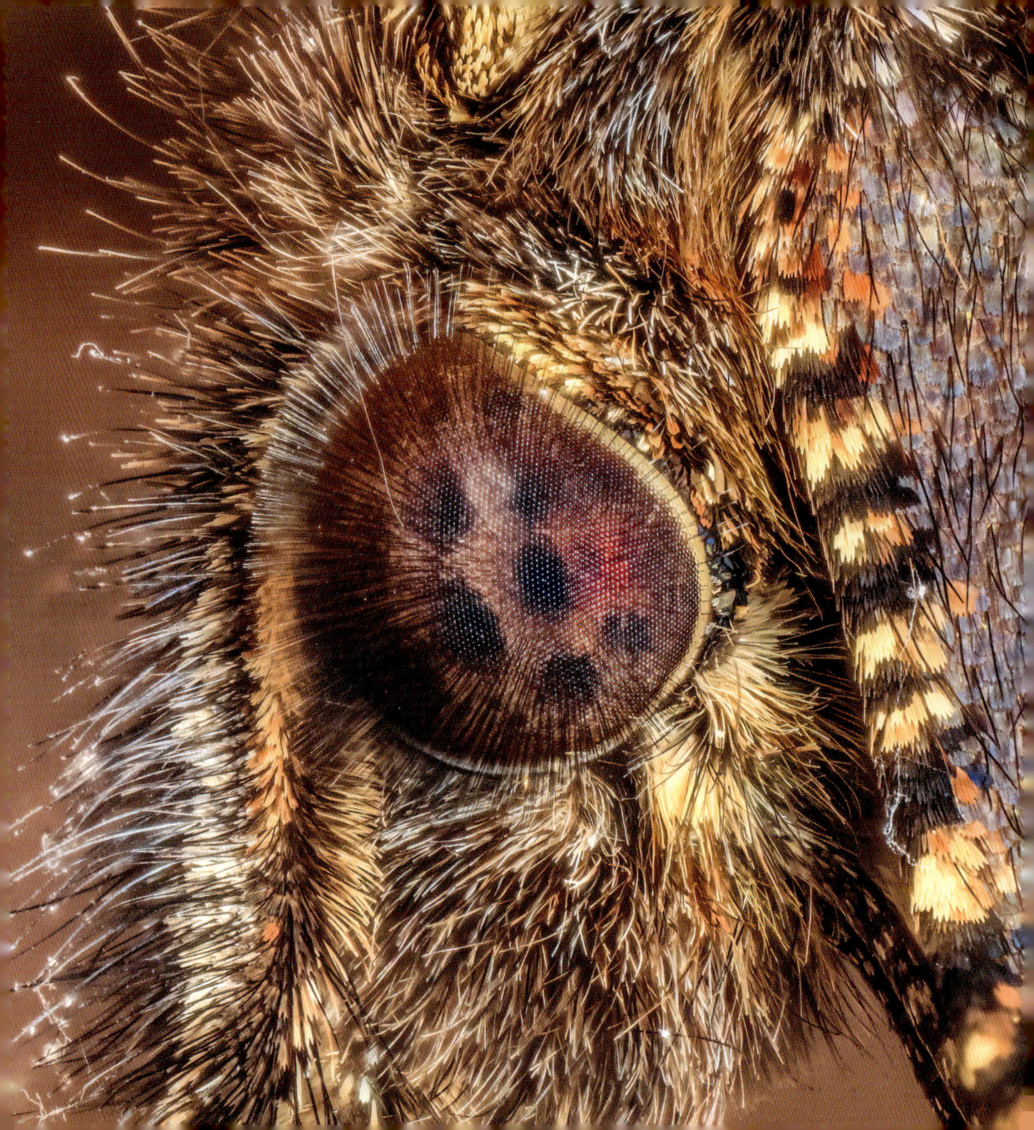

They come in when it is hot outside and our houses seem like a cool sheltering alternative. Once our central heating is turned up against really cold conditions outside, they wake and have nowhere to go and no nectar to be had outside.

BC suggests that on finding a flying, disorientated butterfly indoors in the winter, the first thing to do is calm it in a cardboard box placed in a cool place for about 30 minutes. It could then be placed into an unheated shed/porch/garage, always remembering that it will need to be able to leave when it reawakens in early spring. Failing this, keep the butterfly cool indoors and ready to release once the weather warms up.

Migration

Our window on butterfly migration, which has been called 'one of the most spectacular natural phenomena in the world' is provided by Rothamsted Research's two upward-facing radars in Hampshire and Hertfordshire. It had been known since the mid-1800s that 'clouds' of butterflies could cross the Channel. Observers during the Second World War even spotted a golden ball above the water that turned out to be a mass migration of yellow butterflies. Rothamsted confirmed that these well documented sightings were not as unlikely as was first thought.

Butterflies like the Painted Lady and the Clouded Yellow were seen coming into the country (most notably, the 2009 Painted Lady invasion) but were never seen to leave. Did this mean that the UK was a dead end for migrant butterflies? It would seem not. Rothamsted's radars showed that the butterflies rose to an average of 500 metres and able to take advantage of the prevailing southerly winds, subsequently travelling south at 30mph.

Left: The eye of a dormant Peacock.

The approximately 12,000km migration to us begins in the spring and starts with a caterpillar in the Sahel region of Africa. This is not usually a single generation move. Once they have flown over the Sahara some of that first group will stop on the way to breed the next generation, who will continue the journey crossing Europe, possibly through another breeding cycle and then finally to the UK. Once the weather warms in early summer subsequent insects will fly the Sahara and, research has shown, will make it here in one flight, whilst others take multiple generations.

It would appear that the biological drive to fly north or south is down to the length of the days. Painted Lady caterpillars experiencing growth during longer days know they should fly north, whilst shorter days signal a southerly flight. Global warming may well change the norms of breeding cycles, with species rarely seen here establishing themselves and regularly overwintering as our coldest months grow milder.

Peacock flight sequence.

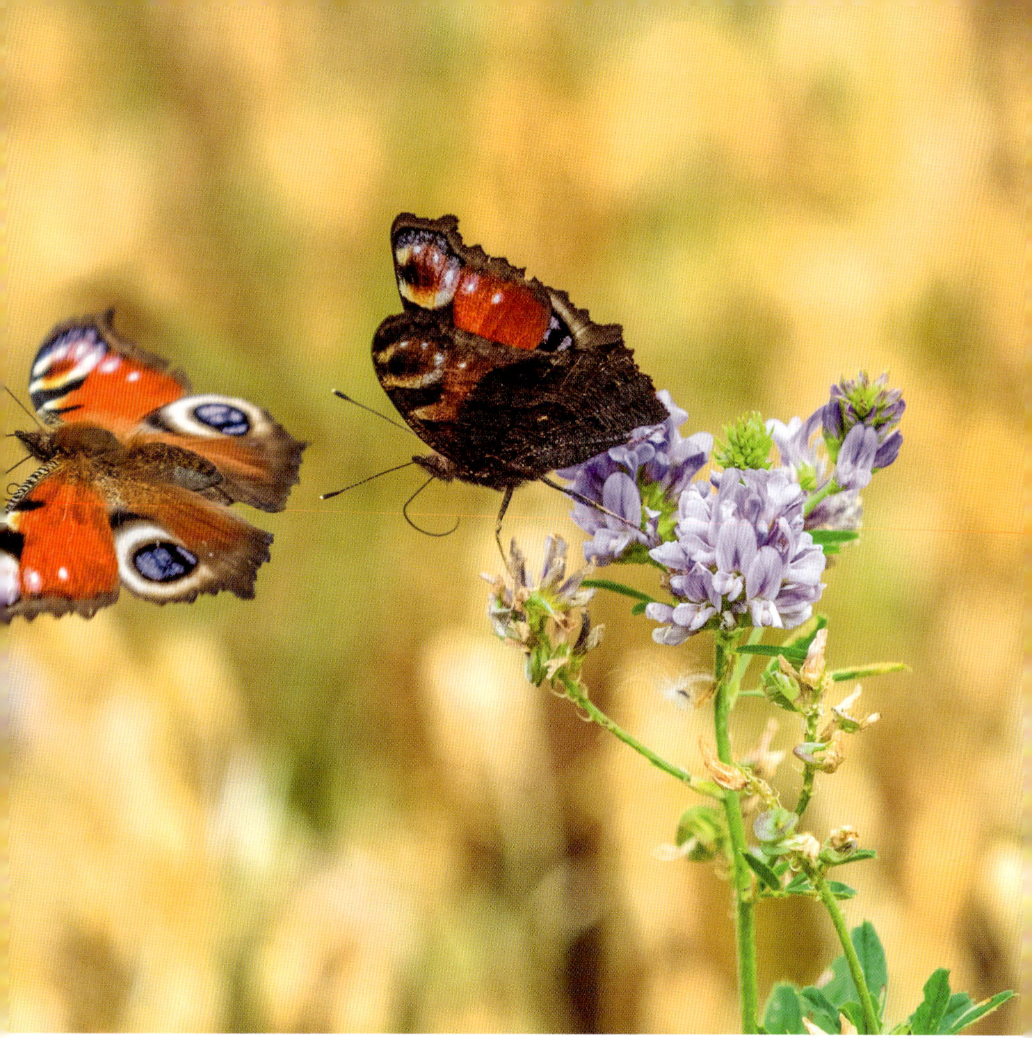

Green Hairstreak flight sequence.

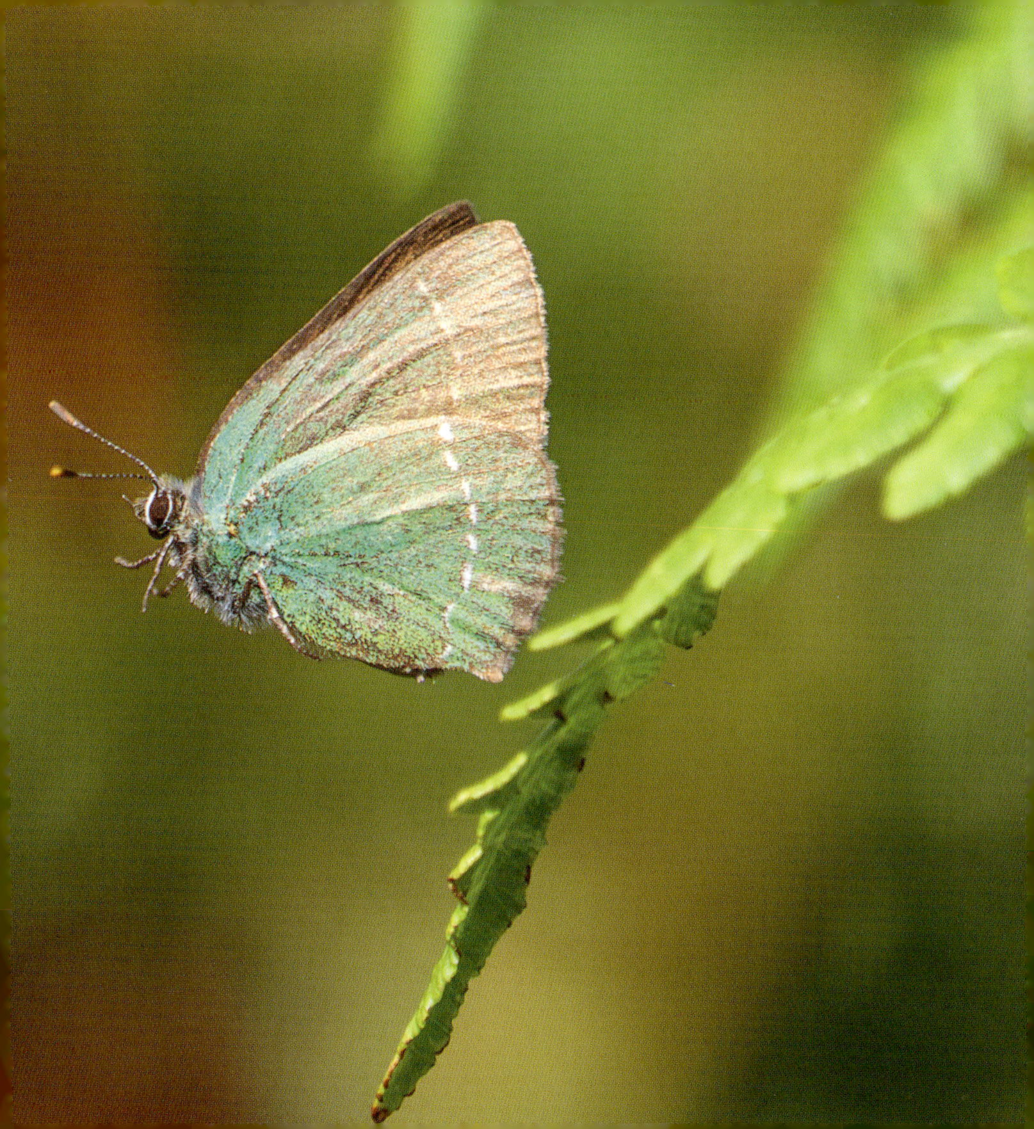

Spotted flycatcher with Red Admiral.

Butterfly Predators and How to Avoid Them

Butterfly Predators and How to Avoid Them

Perhaps we need to begin by looking at how butterflies avoid being eaten!

- Although they do not cause the predator a problem, some are put off by the sheer brightness of the butterfly's colour and patterning. A rough rule of thumb in the natural world is that brightly coloured things are often not good to eat and some butterflies survive on this theory.

- Another strategy for survival is to have a speedy, flitting and erratic flight path to escape.

- Camouflage is used by some butterflies. Pretending to be a dead leaf or twig can keep them safe from predators.

- Butterflies are good at hiding too, (as any serious lepidopterist will tell you.) In amongst trees, under leaves, squeezed in the bark or even in low scrub and grasses.

If you fail with these strategies, you might be eaten by any one of the following:

Birds

Birds catching and eating butterflies may seem wrong but in fact it helps maintain the natural balance of nature.

Warblers, sparrows, flycatchers, robins, thrushes and wrens are all known to eat butterflies amongst other things. Warblers and flycatchers are both adept at catching insects in flight, whilst sparrows in particular are keen feeders on caterpillars. This later demonstrating the constant balance when hunting for prey known as the 'effort to benefit equation';

Dragonfly eating butterfly.

a caterpillar can't move nearly as fast as an adult butterfly.

Dragonflies

Butterflies have met their aerial match in the dragonfly, who are experts at in-flight capture. Fast and manoeuvrable, most dragonflies can outperform a butterfly with ease. The long forelegs capture the prey, which will then be carried to a perch to be consumed.

Bats

The amazing pinpoint accuracy of the bat's echolocation system, means it is the only predator to hunt butterflies in complete darkness and hunt at night. We know by the detritus found at their roosts that long-eared bats in particular are fond of Small Tortoiseshell and Peacock butterflies.

Frog eating butterfly.

Frogs

With the supreme capture tool of a long sticky tongue, frogs can sit and wait for something to go past and gather it up in an instant. Provided the prey will fit in their mouths, they will eat it; spiders, grasshoppers, butterflies and many other insects.

Spiders

The ultimate predating carnivore, we have all seen a butterfly caught in a web. Once entrapped, the spider moves in and immobilises its prey by injecting venom, wrapping up the immobilised insect and leaving to eat later.

Crab Spider eating a butterfly.

Silver-washed Fritillary
with torn wings.

Butterfly Conservation

Butterfly Conservation

Butterfly Conservation is the only UK charity dedicated to saving butterflies, moths and our environment. Their vision is a world where butterflies and moths thrive and can be enjoyed by everyone, forever. Nationally, there are over 40,000 members, making it the largest insect conservation body in all of Europe.

Butterflies and moths are a vital part of our wildlife heritage and are sensitive indicators of the health of our environment. The stark fact is that the numbers of butterflies and moths continue to diminish at an alarming rate. Their data shows they are decreasing faster than most other well documented groups of plants and animals.

Over two-thirds of our resident butterflies are declining and desperately need our help.

Five species have actually been lost altogether to England and Wales in the last 150 years and many others are now found only in small areas where habitat remains. However, Butterfly Conservation is making a difference; without their recording and monitoring data, their science and research and their conservation delivery programmes, the fate of these beautiful insects would be much worse.

Two male Scotch Argus.

A large portion of their work is carried out by dedicated volunteers who are at the heart of Butterfly Conservation. Volunteers enable them to make a greater impact and sustain projects in a way that staff alone could not do. There are 32 branches across the UK, who survey and monitor local sites, undertake practical conservation work and organise events to promote awareness of the difficulties facing our butterflies and moths. There are plenty of opportunities to become involved wherever you live. To locate your nearest branch and find out more about the work of Butterfly Conservation, visit their website, www.butterfly-conservation.org/in-your-area. Everyone who joins Butterfly Conservation is automatically given free membership of their local branch, most of which produce a newsletter which is sent out to members in addition to the national magazine, which people receive three times a year.

Connecting with nature for our physical and mental health is important and recent research provides evidence that green spaces and interaction with nature has a positive effect on well-being, making us happier and healthier. Everyone has a part to play, starting in our own gardens. By choosing plants rich in nectar we can attract butterflies and moths to visit and if we encourage some of the plants on which caterpillars feed we can encourage them to stay and breed.

Wild Spaces campaign

As part of its national strategy, Butterfly Conservation has launched its Wild Spaces campaign www.wild-spaces.co.uk where you can find lots of useful tips and advice on how you can turn your garden and other areas of green space into a wildlife haven. Do please join; ultimately what is good for butterflies and moths is also good for us humans.

www.butterfly-conservation.org

Volunteers at Bury Ditches, Shropshire, UK.

Grizzled Skipper.

Swallowtail.

A 3500-year-old Egyptian fresco in Luxor features the oldest illustration of the African Monarch.

Butterflies in Myth and Legend

Butterflies in Myth and Legend

Greek, Roman and Christian

Recognisable Tiger butterflies (*Danaus chrysippus dorippus*) appear on the wall of the Tomb of Nebamun who died nearly 3,500 years ago. They appear larger than they would be in life and, whilst there seems to be no religious implication to them, their size indicates their rank and importance to the Egyptians as an object of beauty in the animal kingdom. Some scholars think there could even be a logical link similarity between the chrysalis and the wrappings of a mummy. The butterfly as a simple symbol of rebirth would not have been lost on the Egyptians of this period. They saw, with the arrival of the annual Nile floods, the desert convert from seemingly dead dust to lush green. The Tiger butterfly can survive across a variety of habitats and this resilience may be why, combined with the natural 'rebirth' of the landscape, they were included in funereal art but are not recognised as a God in the ancient Egyptian world.

The early concept of the soul is embodied in the formal Greek word for a butterfly – Psyche – meaning both the soul and the insect. The idea of the dead looking chrysalis giving birth to a beautiful butterfly is paralleled by the idea of the butterfly bearing away the soul of a man on his death. The Romans also used the term psyche, although the actual Latin translation for soul is *anima*. Psyche the Greek goddess can often be found depicted as a woman with butterfly wings. She was granted immortality by the Gods and given multiple trials to gain the right to be with her inamorata, Eros (Cupid) who was the god of love and desire. This long and winding tale is documented in books 4 to 6 of the 11 volumes of

Psyche Abandoned is an 1819 white marble sculpture
by the Italian sculptor Pietro Tenerani (1789-1869).

The *Golden Ass* by Apuleius in the 2nd century.

With its core essence of resurrection, the butterfly was a Christian symbol too and there is none better, from seeming death (chrysalis) comes soaring life, surely every Christian's hope for rebirth. Before the printing press was invented, copies of the Bible were done in manuscript form by hand. It is now well accepted that to add light relief, bored copyists would sometimes add little drawings in the margins. Marginalia found in many medieval manuscripts included animals of all types and butterflies were no exception.

Above: Archer shooting butterfly in a medieval manuscript.

Above: Catching a butterfly, medieval manuscript.

Irish folklore

Like the psyche of Greek myth, butterflies in Irish folklore are believed to be connected to the soul of humans. It is bad luck to kill a white butterfly as they are thought to be the souls of dead children, whilst the appearance of a yellow butterfly brings reassurance that those who have died are resting happily in heaven. Butterflies are said to be able to pass through the barrier between this world and the next with ease and as such should always be treated with respect.

The Celtic celebration of Beltane and the butterfly are often linked by fire; Beltane is after all the celebration of the coming of summer and bonfires are lit as part of the festivities. The Gaelic word for butterfly is *dean-dhe*, 'light of the Gods'. However, there are other meanings for this name, possibly a magic flame in the need-fire or the Beltane Balefire. Féileacáin can also mean butterflies (literally, 'little flying creature, butterfly') in Irish mythology; the spirits of the dead come back to reassure the living that they are fine.

It is hard to definitively confirm whether the concept of the butterfly soul carrier existed in Irish culture before the arrival of Christianity or whether it was created to assist in the explanation of the soul going to 'heaven'. Apparently there is still an Irish expression *'na féileacán a bhrú as duine'* which translates as to push the butterflies out of a person i.e. to let the soul go.

Japanese and Chinese cultures

The concept of transformation as described by the life-cycle of the butterfly is ever present in both Japanese and Chinese cultures. In China the transition shows strength and a kind of freedom, a new form of living. Like the Celtic thought on this subject, in Japanese society the butterfly is a guide who helps the 'soul' or 'spirit' move to the afterlife. Even today, in traditional Japanese dress, the obi or sash worn over a kimono can be tied into a 'butterfly knot – Chōchō musubi' (below).

Mesoamerica

There is one area of the world where the myth and legend of the butterfly is a centrepiece. Mesoamerica. Derived from the Greek and meaning 'Middle America', it refers both geographically and culturally to a modern-day area covering Mexico, Guatemala, Belize, Honduras, Nicaragua, Costa Rica, and El Salvador. Whilst coming from diverse civilizations: Olmec, Maya, Toltec, Zapotec, and Aztec, they all share cultural characteristics, which include butterflies in their belief systems.

The central core of ancient understanding of the importance of butterflies is that each one carried the soul of a fallen warrior. The origin of this thought is believed to be the large multi-ethnic city of Teotihuacan 100 BC–650 AD. Many murals, artefacts and pots have been found on the site in which butterflies feature. Incense burners in particular were heavily decorated with the insects and there are many of them, supporting the theory offered by some archaeologists that there was a cult of warrior/ butterfly followers.

The Maya people

The Maya people (250–900 AD) took the butterfly to their hearts at the early stages of their civilisation. They observed the life cycle and showed great respect to 'the souls', disrespect meant your ancestors might seek vengeance on you and your loved ones. Mayans also followed Hunab Ku (above) or the 'Galactic Butterfly' or 'One God'. Maya Scholars thought that this Galactic Butterfly represented not just the souls of the dead but the consciousness of all living things, animals, plants and people. The Maya as a people no longer really exist but interestingly, in Central America and other tropical areas, they still today have a close connection to butterflies and nature. The Maya are butterfly farmers, supplying zoos and live exhibitions with pupae. It has become a sustainable industry, depending as it does on the natural vegetation that the insects feed on, so, rather than cutting down their forests, they conserve them.

Above: Guatemala incense burner circa 300–600 C.E.

Left: Tula figures with a butterfly-shaped chest plate.
Below: Aztec butterfly stamp.

Note the stylised butterfly nose decoration on the head at the centre of this incense burner. It is from Guatemala in roughly the same time period as Teotihuacan but the site where it was found is nearly 700 miles from the city. Furthermore, at Tula where the Toltec people built their capital more than 200 years after the fall of Teotihuacan, the symbols and signs they borrowed from that civilisation are plentiful. The image above shows a row of statues on the top of a pyramid, they are armoured warriors with stylised butterfly breastplates so the Toltec knew of the butterfly connection too. Even later the Aztec (1300–1520)

Monarch butterfly migration.

society was still using the butterfly on incense burners and making clay stamps for easy reproduction of the stylised image on fabrics or pottery or even skin.

There is a continuous connection with butterflies over 1,500 years but after the arrival of the Conquistadors and Christianity, the link between the souls of the dead and butterflies faded. However, the forest communities retained their interest in the annual mass migration of the Monarch butterfly.

The migration is just immense and as Mesoamericanist Jesper Nielsen of the University of Copenhagen says of the sheer numbers, "Seeing that many living beings at one time, and such beautiful ones, it affects you. We have to be careful and remember that we aren't ancient Mesoamericans and can't see things as they did. But if one did believe that butterflies could be linked to dead souls, to all those who came before us – this huge anonymous family that we ourselves will someday be

Butterflies in Myth and Legend   123

Left: Day of the Dead celebrations monarch costume.

Right: Monarch skull.

This idea of the butterfly as the centre of your world took hold thousands of years ago and still has echoes today.

NB: I could devote a whole book to the butterfly in Mesoamerican culture but I don't have the space. However, I hope it has inspired you to get researching yourself. It is a simply fascinating subject.

The meaning of colour on butterflies

yellow – joy and success
brown – security and genuineness
white – the soul of a child
white – positivity and stability
white – bad luck
red – passion, and wrath
orange – delight, creativity, and sensuality
green – fertility, and growth
blue – devotion
purple – tranquillity
black – life-altering transformation

part of—the experience would be overwhelming." A resurgence of interest in the Mexican Day of the Dead (DotD) celebrations has meant that current generations are coming new to the importance of butterflies. Core to this interest is the inclusion of the Monarch butterfly into the DotD parades and events. The mass migration has caught the eye of younger generations with no history in this field and street celebrations now include butterfly costumes and butterfly covered skulls.

126   Vincent Van Gogh, 1889, Van Gogh Museum, Amsterdam.

Butterflies in Art and Literature

Butterflies in Art and Literature

Maria Sibylla Merian

Later in this chapter I shall be talking about a famous writer whose interest in entomology began with finding some books in an attic. The books were by Maria Sibylla Merian (1647-1717), one of the most important German entomologists and scientific illustrators of the period. Her work covered both literature and art as not only did she write very knowledgeably about metamorphosis but she drew it in amazing detail as well. Her breakthrough work on the life cycle of insects from egg to adult, because of its careful thoroughness, overturned the previously held theory of spontaneous generation, which asserted that living creatures could develop from non-living matter. Merian explained how she became interested in this area of science: 'I spent my time investigating insects. At the beginning, I started with silkworms in my home town of Frankfurt. I realized that other caterpillars produced beautiful butterflies or moths, and that silkworms did the same. This led me to collect all the caterpillars I could find in order to see how they changed.' Her work was published in 1679, the first of a fully illustrated two-volume set, *The Caterpillars' Marvellous Transformation and Strange Floral Food*, on the

metamorphosis of insects.

From her teens Maria was driven to draw insects from life, this set her apart from those who were drawing from 'specimens'. It was not unusual at that time to illustrate documented work on insects like butterflies and moths. What made her work outstanding was that she was a trained artist and keen to ensure the colour accuracy of her paintings. Engravings were made which were always as faithful as she could make them to the original watercolours which she had created on vellum. Furthermore, her collecting and drawing of the foodstuff of her subjects and her cataloguing of their different stages was unusual and meant that her work often recorded the development of plants from bud to leaf to flower as well.

Before 1690, Merian had an interesting life; an unhappy marriage, two daughters and the need for a stable haven. She and her widowed mother and the daughters

Above: *The Caterpillars' Marvellous Transformation and Strange Floral Food* by Maria Sibylla Merian, 1679.

Metamorphosis Insectorum Surinamensium by Maria Sibylla Merian, 1705.

found just that in the Labadist Community in Wieuwerd, Friesland. It was disciplined, religious and ran as a proper communal community. (The husband Johann Graff was refused entry but tried twice.) Merian had time at last to study natural history and, perhaps most importantly, Latin, to help her read the scientific books of the day.

Following the death of her mother in 1690 she and her girls moved to Amsterdam and two years later she and Johann divorced. Maria was now free, she taught flower painting and sold her work and within eight years she had a comfortable home and a secure life.

However, the marriage of her daughter Johanna in 1692 to a merchant who traded with Suriname opened up the possibility of feeding her insatiable appetite for information about insect transformation and new species. In 1699 she and her youngest daughter Dorothea, set off on a five year adventure to find and record through illustrations, insects in Suriname. To fund this enterprise she sold 255 paintings and this in itself was unusually as most trips of this sort were funded by companies or countries and were undertaken by men. The hundreds of drawings and notes that the two women brought back with them influenced future generations to the present day. Her discoveries included unknown plants and animals and as always, she recorded in both writing and drawings the surrounding habitat, what they fed on and what the indigenous people called them. Her interest spread to the medicinal use of the plants too, a fact which caught the interest of scientists and doctors alike.

Three years after her return, following encouragement from the many visitors who called to see her outstanding collection, she self-published *Metamorphosis insectorum Surinamensium*, writing about it 'Now that I had returned to Holland and several nature-lovers had seen my drawings, they pressured me eagerly to have them printed. They were of the opinion that this was the first and most unusual work ever painted in America.'

It cannot be overemphasised how important her life's work was in opening up the understanding of new species of all types. Merian was so ahead of her time in her attitude to recording the colours, detail and specifics of so many different insects. All were observed and drawn from life, not from dead samples as was done at the time. Over the decades she has gathered a vast cohort of admirers, including Sir David Attenborough. It is hoped that this brief insight into her life will spur you to research the life and work of Maria Sibylla Merian yourself, it is worth the effort.

"Saw the children of earth & the tenants of air." p.4

The Butterfly's Ball and The Grasshopper's Feast by William Roscoe

The Butterfly's Ball and The Grasshopper's Feast is by William Roscoe (1783–1831) who is perhaps best known as one of the UKs first abolitionists but also an historian, Christian philanthropist, founder of Liverpool Botanic Garden and poet.

Above: Illustration from *The Butterfly's Ball and The Grasshopper's Feast* by William Roscoe.

The Butterfly's Ball and The Grasshopper's Feast was written for his youngest son Robert, the little Robert who appears throughout the poem, which describes a collection of small animals and insects setting out in a jolly throng for a ball in the woods. It is a charming evocation of friends enjoying an evening out, it is however rather long but here are the first two verses... they all have a good time and go home.

Come take up your Hats,
and away let us haste
To the Butterfly's Ball, and the
Grasshopper's Feast.
The Trumpeter, Gad-fly,
has summon'd the Crew,
And the Revels are now only waiting
for you.
So said little Robert, and pacing along,
His merry Companions came forth
in a Throng.
And on the smooth Grass,
by the side of a Wood,
Beneath a broad Oak that for Ages
had stood, [...]

Alan Aldridge

More familiar to the modern reader is *The Butterfly Ball and the Grasshopper's Feast* by Alan Aldridge (illustration), William Plomer (verse) and Richard Fitter (nature notes), published in 1973. Aldridge was a very well known graphic designer with his own firm, INK which worked for many musicians on album covers in the 1960s and 70s. He had a wonderfully exuberant, vigorous, swirling style that takes you into each image. This version, whilst based on the Roscoe original, is more interested in the animals individually and how they prepare for the ball. The was also a subsequent concept album by Roger Glover of Deep Purple.

Alan Aldridge with posters and paintings in Bijenkorf.

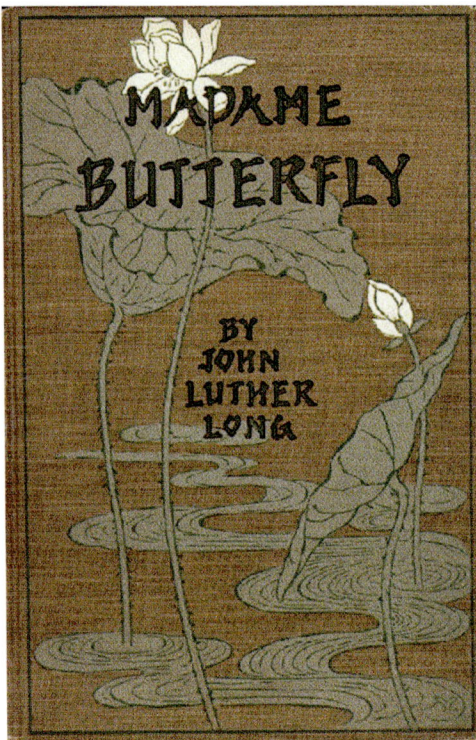

Above: Cover of the original short story of *Madame Butterfly* by John Luther Long, 1903.

Madame Butterfly by John Luther Long

The short story *Madame Butterfly* was inspired by the stories told to John Luther Long by his sister Jennie Correll. Jennie and her husband had been Nagasaki based missionaries in Japan and she explained the culture of the unusual place to him. There was a distinct belief in the late 1880s of something known as a temporary or 'Japanese marriage', where a foreigner would make a marriage of convenience so that he could co-habit whilst he was away from home. Abandoned Japanese women were not unusual at this time. A Japanese and non-Japanese marriage was not permitted by law until 1873 and the Japanese side of the pair would have to forgo their nationality and any social standing. With this rigid and formal society, where love was thought to be inessential to marriage as a backdrop, Correll told her brother the story of a "dear little teahouse girl" called Cho-San who was abandoned by the father of her child and it caught his imagination. (NB: in Japanese the word *chou* and the word *chouchou* both mean butterfly and are interchangeable.) Educated in Europe where he became a barrister, back in Japan Baron Hozumi proposed a law in 1898 that the absence of love should be made a ground for divorce. It failed to pass.

In Long's story Cho-Cho-San was very young – 17, beautiful and, to the American Navy Lieutenant Benjamin Franklin Pinkerton, exotic, just like the butterfly her name conjured. Pinkerton marries her, makes her pregnant and then departs promising to return 'when the robins nest again'. Three years later, when she has been shunned by her family and the community and despite a good offer of marriage, Cho-Cho-San is still awaiting his return.

One day she sees his ship in the harbour and with the help of her maid Suzuki, she prepares the house and, wearing her finest kimono, she and her son Trouble sit to watch for his arrival. He does not come. A week later however, a steamship docks and Cho-Cho-San sees that Pinkerton is on board with a blonde woman, who she subsequently finds out is Pinkerton's wife Kate.

Cho-Cho-San learns that Kate intends to plead with her to take Trouble back to American with them, so she says her farewells to Suzuki and Trouble and goes to her room to commit suicide. The attempt fails and Suzuki dresses the wound Cho-Cho-San has made and when Kate gets to the house she finds it abandoned.

Although when printed in the *Century Magazine* it was not an instant success, Long's publishers put it together with some of his other stories in a book and it sold very well indeed. So much so that they produced an illustrated reprint in 1903 (see left).

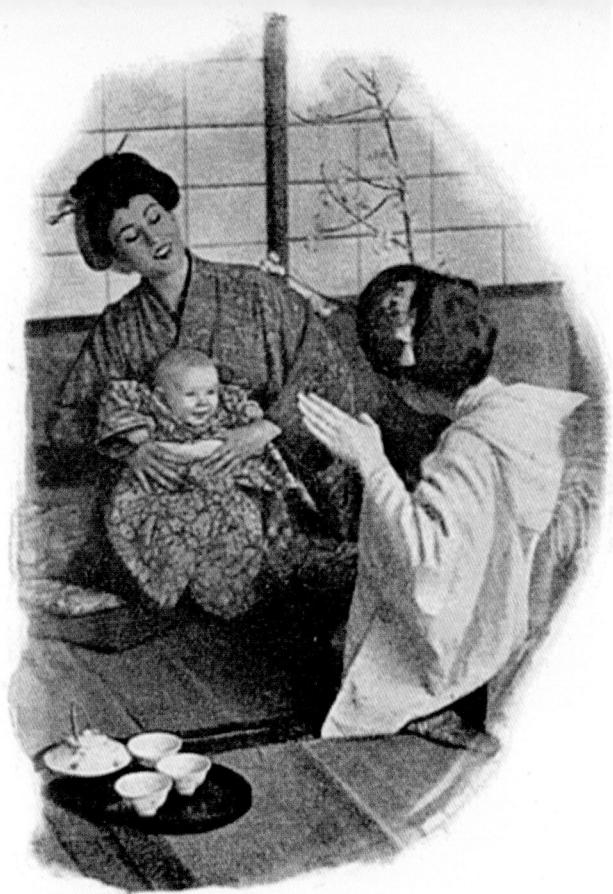

" But—he is a miracle ! Yes ! "

Above: Madame Butterfly, 1903.

One Act Play: (1900) *Madame Butterfly: A Tragedy of Japan* by David Belasco and John Luther Long

A collaboration between John Luther Long and the successful playwright of the day David Belasco, (*The Girl Of The Golden West*) resulted in a one act stage play based on Long's short story, with one major change. At the end of the play Pinkerton visits the little house with Kate to see if they can take the child Trouble back to America with them. However, they are too late and Cho-Cho-San has already cut her throat and this time dies, breathing her last in the arms

of Pinkerton saying "Too bad those robins didn' nes' again."

It was a huge success for Belasco, opening for a seven week run in New York, followed by packed houses at the Duke of York's in London. The first Cho-Cho-San was played in New York by Blanche Bates and the first Pinkerton by the British/American actor Frank Worthing.

When the play transferred to London in the summer of 1900, Giacomo Puccini was in the audience and was captivated by the story.

Above: Blanche Bates and Frank Worthing.

Solomiya Krushelnytska
as Butterfly, 1904.

Opera: (1904) *Madama Butterfly* by Giacomo Puccini, libretto by Luigi Illica & Giuseppe Giacosa

The opera barely deviates from the play, except in the spelling of the name Cio-Cio-San and the fact that unbeknown to Pinkerton, Butterfly has converted to Christianity before they are married. The ceremony is at the house and afterwards, an uninvited uncle arrives to curse her conversion and order all the guests to leave, which they do, cursing her too. The rest of the story follows the play and she dies at the end as Pinkerton realises he has made a terrible mistake.

Between 1904 and 1907 Puccini wrote five versions of *Madama Butterfly.* He was enthralled by the story and researched Japanese folk songs, wanting to show a real version of Japan.

February, 1904: The world premiere of the original two-act version at La Scala Milan. It was a disaster, probably due to under rehearsal and constant edits.

May, 1904: Puccini rewrote it in three acts and it was a huge success with Solomiya Krushelnytska as Cio-Cio-San. This is the second version, which premiered in the States in 1906.

1906: A third version was completed and performed at the Met in New York in 1907.

1907: Puccini made edits to the orchestral and vocal scores of the third version to make the fourth version.

1907: The final revisions were made and this became the fifth version, which is usually performed today.

This is now known as the 'Standard Version'.

In spite of the shift in attitudes to depictions of women/cultures not indigenous to the singers/marriage/ society in general, *Madama Butterfly* remains one of the most popular operas of Puccini's canon.

Above: Nevill Holt Opera Festival, 2010, costumes by Mark Bailey.

Peonies and Butterfly, Hokusai's Large Flowers series.

Chuang Tzu and the Butterfly by Li Po (701–762)

Chuang Tzu in dream became a butterfly,
And the butterfly became Chuang Tzu at waking.
Which was the real—the butterfly or the man?
Who can tell the end of the endless changes of things?
The water that flows into the depth of the distant sea
Returns anon to the shallows of a transparent stream.
The man, raising melons outside the green gate of the city,
Was once the Prince of the East Hill.
So must rank and riches vanish.
You know it, still you toil and toil,—what for?

A very philosophical question from the ancient philosopher
Chuang Tzu inspired this poem. He dreamed he was a butterfly and
on waking asked himself the following question: 'Was I Chuang Tzu
dreaming I was a butterfly or am I now really a butterfly dreaming
that I am Chuang Tzu?' Philosophers ponder it still.

Brimstone in flight.

Arakida Moritake (1473–1549)

The falling flower
That drifted back to the branch
Was a butterfly

John Ray (1617–1705)

You ask what is the use of
butterflies?
I reply to adorn the world and delight
the eyes of men;
to brighten the countryside like so
many golden jewels.
To contemplate their exquisite beauty
and variety is to
experience the truest pleasure.

Above: Japanese butterflies and
moths.
Right: Wall on Ragwort.

To a Butterfly by William Wordsworth (1770–1850)

I've watched you now a full half-hour;
Self-poised upon that yellow flower
And, little Butterfly! indeed
I know not if you sleep or feed.
How motionless!--not frozen seas
More motionless! and then
What joy awaits you, when the breeze
Hath found you out among the trees,
And calls you forth again!

This plot of orchard-ground is ours;
My trees they are, my Sister's flowers;
Here rest your wings when they are weary;
Here lodge as in a sanctuary!
Come often to us, fear no wrong;
Sit near us on the bough!
We'll talk of sunshine and of song,
And summer days, when we were young;
Sweet childish days, that were as long
As twenty days are now.

Top: Scotch Argus.

STAY near me--do not take thy flight!
A little longer stay in sight!
Much converse do I find in thee,
Historian of my infancy!
Float near me; do not yet depart!
Dead times revive in thee:
Thou bring'st, gay creature as thou art!
A solemn image to my heart,
My father's family!

Oh! pleasant, pleasant were the days,
The time, when, in our childish plays,
My sister Emmeline and I
Together chased the butterfly!
A very hunter did I rush
Upon the prey:--with leaps and springs
I followed on from brake to bush;
But she, God love her, feared to brush
The dust from off its wings.

Camberwell Bathhouse.

From *British Butterflies* (1908), edited by Edward Thomas, from the chapter 'Some English Butterflies' by Anthony Collett

'When the first day comes in March when the air is quick with awakening life, and the earth drinks deep of new, hot, golden splendour from a sun now high in heaven, the seal is set on returning spring by the great yellow wing of the Brimstone butterfly purposefully beating down the rides and lanes like a visible concentration of the light.

With him, or even before him, in the illusory brightness of some halcyon winter noon, there appear three or four other species of a different family, of which the characteristic predominant colour is deep and brilliant red. The commonest of these early spring butterflies are the Small Tortoiseshell, the Peacock with his rich eye pattern and the Brimstones, male and female, in their brilliant yellow and delicate primrose-green.'

This mosaic on the end of the Passmore Edwards Library and Bathhouse by Burgess Park in Camberwell, London, was rescued in 1982 from the front of the Samuel Jones paper factory on Southampton Way. The butterfly art lives on.

Above: Small Tortoiseshell by Kate Kekwick. Drawing, to paper cut-out, then colour is added digitally.

Clouded yellow.

The Butterfly upon the Sky by Emily Dickinson (1830–1886)

The Butterfly upon the Sky,
That doesn't know its Name
And hasn't any tax to pay
And hasn't any Home
Is just as high as you and I,
And higher, I believe,
So soar away and never sigh
And that's the way to grieve—

Symphony in Yellow by Oscar Wilde (1854-1900)

An omnibus across the bridge
Crawls like a yellow butterfly,
And, here and there a passer-by
Shows like a little restless midge.

Big barges full of yellow hay
Are moored against the shadowy wharf,
And, like a yellow silken scarf,
The thick fog hangs along the quay.

The yellow leaves begin to fade
And flutter from the temple elms,
And at my feet the pale green Thames
Lies like a rod of rippled jade.

Harpsicohord painted by Nancy Jones.

Butterfly Harpsichord by Jane Toole

For Mike Bardsley, harpsichordist extraordinaire

A flash of crisp colour
No sustain to remain
Transient beauty.
Bach on a path to the green hedgerow.
Fast crisp notes, darting kaleidoscope
Skipping, skimming flight on flower and key
The power to delight in flight might just catch you unaware.

Vladimir Nabokov

In a 1967 interview, a famous author told the *Paris Review*, 'The pleasures and rewards of literary inspiration are nothing beside the rapture of discovering a new organ under the microscope or an undescribed species on a mountainside in Iran or Peru. It is not improbable that had there been no revolution in Russia, I would have devoted myself entirely to lepidopterology and never written any novels at all.' You may be surprised to learn that that famous author was Vladimir Nabokov (1899–1977), who continuously during his literary professorships and writing was to be found in the laboratories of both Harvard as a research fellow in zoology and de facto curator of the butterfly collection at the university's Museum of Comparative Zoology in the 1940s and Cornell in the 1950s.

What ignited Nabokov's curiosity in butterflies was finding some books by Maria Sibylla Merian, the entomologist and scientific illustrator, in the attic of the family country house in Vyra and so began a lifelong passion. Every summer in the western U.S. he would take a butterfly collection trip, always accompanied by his wife, Vera Slonim, out of necessity, as he had never learned to drive.

In honour of his work on the Polymmatini of family *Lycaenidae*, the genus *Nabokovia* was named after

Spring Azure (*Celastrina ladon*) butterfly.

him. During his life he was rather looked down on by professional lepidopterists, perhaps because he was so successful in his other career. However, genetic research in January 2011 confirmed the hypothesis he had offered up in 1945: that a butterfly species called Polyommatus Blues arrived in America from Asia over millennia in a series of waves. Gene sequencing at Harvard has proved that there were in fact five waves of butterflies and in February 2011 a paper was given at the Royal Society in which Nabokov was confirmed as being absolutely correct in this hypothesis.

Whilst fans of both his literary and his scientific skills have tried to show that one influenced the other, perhaps the truth was that as he once said himself, 'A writer should have the precision of a poet and the imagination of a scientist'.

Grayling take-off.

The Example by William H. Davies (1871–1940)

Here's an example from
A Butterfly;
That on a rough, hard rock
Happy can lie;
Friendless and all alone
On this unsweetened stone.

Now let my bed be hard,
No care take I;
I'll make my joy like this
Small Butterfly;
Whose happy heart has power
To make a stone a flower.

The Very Hungry Caterpillar
by Eric Carle

Anyone with a child born after 1969, when this book was first published, must have sampled the delights of *The Very Hungry Caterpillar*, Eric Carle's tale of metamorphoses. Eating an extra item a day (one apple on Monday, two pears on Tuesday), he is forever growing and yet still hungry. The trick of the book is that the caterpillar eats through each page to get to the next – an idea which came to Carle when he was using a two-hole punch and wondered if it would be possible to do it in a book.

Finally, the caterpillar, after stuffing himself on the final day with ten different foods, including ice cream and a sausage, builds a cocoon, and two weeks later a beautiful butterfly emerges. It is unfortunate that this was a rather glaring mistake, as you will know if you have read the chapter about the life cycle.

A cocoon produces a moth, whereas a butterfly comes out of a chrysalis.

That said, the title has sold over 50 million copies and been translated into 60 languages, with the first generation of children it delighted now reading it to their grandchildren and great-grandchildren.

Adonis Blue in flight.

Blue-Butterfly Day by Robert Frost (1874–1963)

It is blue-butterfly day here in spring,
And with these sky-flakes down in flurry on flurry
There is more unmixed colour on the wing
Than flowers will show for days unless they hurry.
But these are flowers that fly and all but sing:
And now from having ridden out desire
They lie closed over in the wind and cling
Where wheels have freshly sliced the April mire.

Butterfly by D. H. Lawrence
(1885–1930)

Butterfly, the wind blows sea-ward,
    strong beyond the garden-wall!
Butterfly, why do you settle on my
    shoe, and sip the dirt on my shoe,
Lifting your veined wings, lifting them?
    big white butterfly!

Already it is October, and the wind
    blows strong to the sea
from the hills where snow must have
    fallen, the wind is polished with
       snow.
Here in the garden, with red
    geraniums, it is warm, it is warm
but the wind blows strong to sea-ward,
    white butterfly, content on my shoe!

Will you go, will you go from my warm house?
Will you climb on your big soft wings,
    black-dotted,
as up an invisible rainbow, an arch
till the wind slides you sheer from the
    arch-crest
and in a strange level fluttering you go
    out to sea-ward, white speck!

Male Large Blue.

Butterfly Laughter by Katherine Mansfield (1888–1923)

In the middle of our porridge plates
There was a blue butterfly painted
And each morning we tried who should reach the butterfly first.
Then the Grandmother said: 'Do not eat the poor butterfly.'
That made us laugh.
Always she said it and always it started us laughing.
It seemed such a sweet little joke.
I was certain that one fine morning
The butterfly would fly out of our plates,
Laughing the teeniest laugh in the world,
And perch on the Grandmother's lap.

The Butterfly by Pavel Friedmann (1921–1944)

The last, the very last,
So richly, brightly, dazzlingly yellow.
Perhaps if the sun's tears would sing
against a white stone...

Such, such a yellow
Is carried lightly way up high
It went away I'm sure because it
wished to
kiss the world goodbye.

For seven weeks I've lived in here,
Penned up inside this ghetto
but I have found my people here
The dandelions call to me
And the white chestnut candles in
the court.
Only I never saw another butterfly.

That butterfly was the last one,
Butterflies don't live in here,
In the ghetto.

The poet Pavel Friedmann was a Czechoslovakian poet and this poem was written in the Prague Ghetto, Theresienstadt in 1942. It was subsequently found amongst a cache of papers after the war. He died in the Nazi concentration camp at Auschwitz in 1944. Friedmann became famous for this poem after his death not only for its sentiments but also for the many writings and events it inspired. For instance, The Butterfly Project at the Holocaust Museum Houston, caused 1.5 million paper butterflies to be made to symbolise the number of children murdered by the Nazis in the Holocaust.

Gatekeeper in flight.

The Gatekeeper, from *Communion*
by Deborah Harvey (IDP)

There's no one left to mourn them.
Only stony-faced angels keep watch
over the names of forgotten children
written in lichen, blotted with moss

Flowers must bring themselves
dandelions for Mary Kate
stately cuckoo-pints for Diana, the siren
shine of malevolent berries no longer a worry
From the tower the clock strikes four quarters and one
a gatekeeper settles on a stone
its wings the colours of autumn fallen
umber and rust

A clatter of jackdaws bustling back
tatters the death pall with tender talk
Playground voices shoulder through oak trees
boisterously singing the Hokey-Cokey

Photo credits and artworks

Long-tailed Blue.

Acknowledgements

With thanks as always to Craig Jones and David Williams; may the Three Amigos keep riding out! Thanks to Jenny Joy for Grayling egg laying and Stephen Barlow for Large Heath and Mike Smith for Large Chequered Skipper and Kirsty Gibbs for Long-tailed Blue.

Special thanks to Jane Russ who has uncovered so many interesting facts and stories about butterflies, Bravo!
Andrew Fusek Peters

To my wonderful co-author, whose fabulous photography inspired me to find the words to match, many thanks. Let us hope we finally get to meet up this year.

My husband and chum Mick Toole, who is always supportive and always positively critical, I never take you for granted.
Jane Russ

The Butterfly Book
Published in Great Britain in 2025 by Graffeg Limited.

Written by Andrew Fusek Peters and Jane Russ copyright © 2024. Designed and produced by Graffeg Limited copyright © 2024. Series editor Jane Russ.

Graffeg Limited, 24 Stradey Park Business Centre, Mwrwg Road, Llangennech, Llanelli, Carmarthenshire, SA14 8YP, Wales, UK. Tel: 01554 824000. www.graffeg.com.

Andrew Fusek Peters and Jane Russ are hereby identified as the authors of this work in accordance with section 77 of the Copyright, Designs and Patents Act 1988.

A CIP Catalogue record for this book is available from the British Library.

The publisher gratefully acknowledges the financial support of this book by the Books Council of Wales. www.gwales.com.

Printed in China TT011124

ISBN 9781802586541

1 2 3 4 5 6 7 8 9

MIX
Paper | Supporting responsible forestry
FSC® C016973

Books in the Nature Series

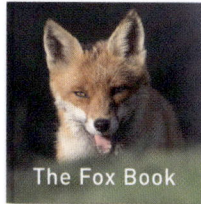

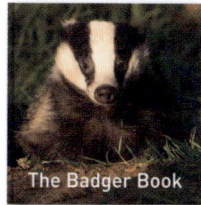

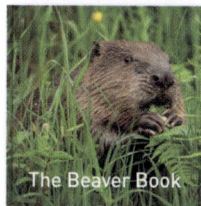

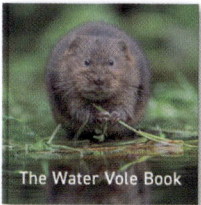

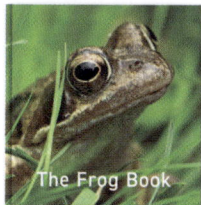

The Hare Book

The Fox Book

The Owl Book

The Red Squirrel Book

The Bee Book

The Robin Book

The Badger Book

The Hedgehog Book

The Native Pony Book

The Puffin Book

The Beaver Book

The Otter Book

The Water Vole Book

The Frog Book

The Crow Family Book

www.graffeg.com

NORTH YORK MOORS

Ordnance Survey

Contents

Walk 1

COOK'S MONUMENT AND ROSEBERRY TOPPING

Distance
4.5 miles / 7.2 km

Time
2¾ hours

GO BY TRAIN

Start/Finish
Gribdale Gate

Parking TS9 6HW
Forestry England Gribdale
Gate parking area

Cafés/pubs
Picnic benches; Great
Ayton

Fine moorland circuit, a national hero's column and a mini Matterhorn

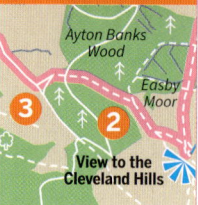

Ayton Banks
Wood

Easby
Moor

3

2

**View to the
Cleveland Hills**

Page 14

Walk 2

OSMOTHERLEY

Distance
3.8 miles / 6.1km

Time
2¼ hours

CATCH A BUS

Start/Finish
Osmotherley

Parking DL6 3AG
Roadside parking in
Osmotherley

Cafés/pubs
Osmotherley

**Attractive village;
lovely reservoir
and Vale of
Mowbray views**

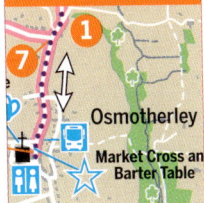

Walk 3

SUTTON BANK AND KILBURN WHITE HORSE

Distance
3.6 miles / 5.8 km

Time
2 hours

Start/Finish
Sutton Bank National
Park Centre

Parking YO7 2EH
Sutton Bank National
Park Centre car park

Cafés/pubs
Picnic benches; Park Life
Café; Gliders' Nest Café

**Best views from
Sutton Bank;
soaring gliders;
white horse hill
figure**

Walk 4

RIEVAULX ABBEY

Distance
2.6 miles / 4.1km

Time
1½ hours

Start/Finish
Rievaulx

Parking YO62 5LB
Village car park

Cafés/pubs
Rievaulx Abbey Café;
Helmsley

**Splendid abbey
ruin; meadows,
woods and
tranquil Ryedale**

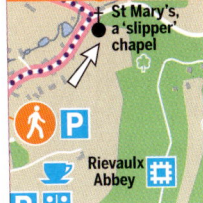

Contents **3**

Walk 5

FARNDALE

Distance
2.9 miles/4.6km

Time
1¾ hours

Start/Finish
Low Mill, Farndale

Parking YO62 7UY
Car park at Low Mill

Cafés/pubs
Daffy Caffy at High
Mill; Farndale Store,
Low Mill

April daffodil
spectacle; Dove
riverside stroll;
pastoral, peaceful
dale

Page 42

Walk 6

HUTTON-LE-HOLE

Distance
4.3 miles/7km

Time
2½ hours

Start/Finish
Hutton-le-Hole

Parking YO62 6UB
National Park car park

Cafés/pubs
Hutton-le-Hole;
Lastingham

Great moorland
views; charming
villages; good
refreshment
options

Page 48

Walk 7

GROSMONT TO GOATHLAND

Distance
3.5 miles/5.5km

Time
2 hours

GO BY TRAIN

Start Grosmont

Finish Goathland

Parking Grosmont,
YO22 5QE; Goathland,
YO22 5LX

Cafés/pubs
Grosmont; Beck Hole;
Goathland

Delightful old
Whitby to
Pickering railway
path; catch a steam
train back

Page 54

Walk 8

WHITBY ABBEY

Distance
3.1 miles/4.9km

Time
1¾ hours

Start/Finish
Whitby Abbey

Parking YO22 4EH
Whitby Abbey
Headland car park

Cafés/pubs
Whitby Brewery;
seasonal ice-cream
van at abbey

Bracing clifftop
stroll; legend
of Dracula and
atmospheric
abbey ruin

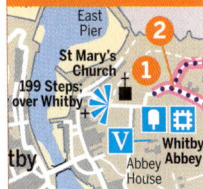

Walk 9

ROBIN HOOD'S BAY

Distance
3.1 miles/5km

Time
1¾ hours

CATCH A BUS

Start/Finish
Robin Hood's Bay

Parking YO22 4RA
Station Road car park

Cafés/pubs
Robin Hood's Bay

Leafy, easy-going
old railway path;
great sea views
on the coast path

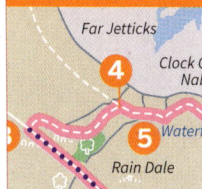

Walk 10

BROXA FOREST

Distance
3.4 miles/5.5km

Time
2 hours

Start/Finish
Broxa Forest

Parking YO13 0JT
Forestry England
Broxa Forest car park

Cafés/pubs
None

Forest drives;
quiet trails; mixed
woodland; open
meadow glades

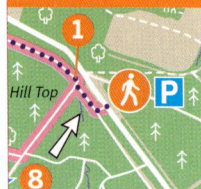

GETTING OUTSIDE IN THE NORTH YORK MOORS

> **The crowning glory of the national park is its extensive heather moorland, which in late summer becomes a beautiful expanse of pink and purple**

OS Champions
Tom and Essi Troughton

Rievaulx Abbey

A very warm welcome to the new Short Walks Made Easy guide to the North York Moors – what a fantastic selection of leisurely walks we have for you!

Covering 554 square miles of wonderful Yorkshire countryside, the North York Moors National Park is an area of upland bounded by the Cleveland plain and Tees Valley to the north, the vales of York and Mowbray to the west, the Vale of Pickering to the south, and the North Sea to the east, ranging from Thirsk to Scarborough. The ten superb walks in this guide explore different characteristics of the landscape and its heritage.

The crowning glory of the national park is its extensive heather moorland, which in late summer becomes a beautiful expanse of pink and purple – pick the walks to Captain Cook's Monument and from Hutton-le-Hole to witness this unforgettable sight. In Farndale, in spring, it is a carpet of yellow daffodils that captivates visitors. The tranquilly set romantic ruin of Rievaulx Abbey, nestled in the pastoral seclusion of Ryedale, contrasts with the haunting Gothic edifice of Whitby Abbey, standing resolutely on its headland hilltop. Magnificent views can be admired from Sutton Bank (James Herriot's 'finest view in England') and from the clifftop vantage point above Robin Hood's Bay. You can immerse yourself in the mixed woodland and open glades of Broxa Forest, catch a steam train at Goathland to complete the walk from Grosmont, and enjoy post-stroll refreshment in charming Osmotherley.

Tom and Essi Troughton, OS Champions

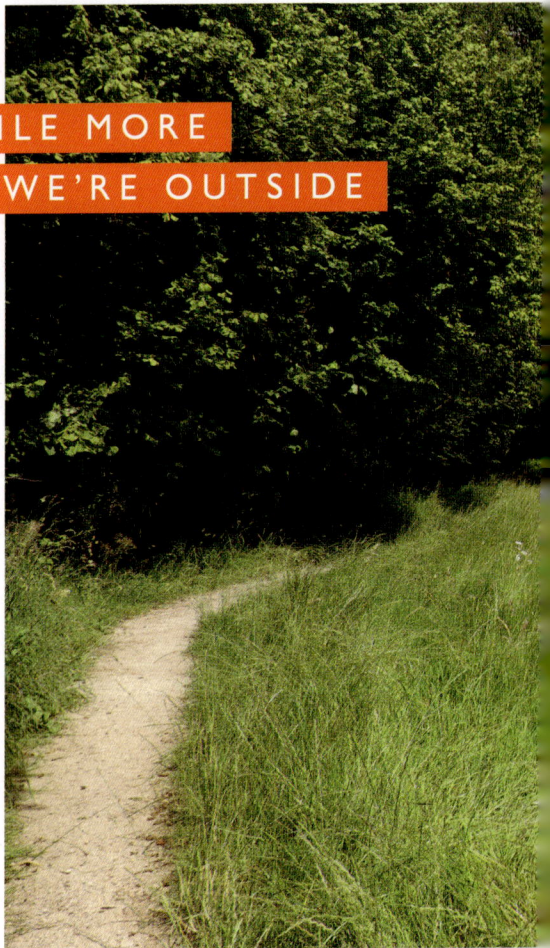

WE SMILE MORE WHEN WE'RE OUTSIDE

Farndale

Whether it's a short walk during our lunch break or a full day's outdoor adventure, we know that a good dose of fresh air is just the tonic we all need.

At Ordnance Survey (OS), we're passionate about helping more people to get outside more often. It sits at the heart of everything we do, and through our products and services, we aim to help you lead an active outdoor lifestyle, so that you can live longer, stay younger and enjoy life more.

We firmly believe the outdoors is for everyone, and we want to help you find the very best Great Britain has to offer. We are blessed with an island that is beautiful and unique, with a rich and varied landscape. There are coastal paths to meander along, woodlands to explore, countryside to roam, and cities to uncover. Our trusted source of inspirational content is bursting with ideas for places to go, things to do and easy beginner's guides on how to get started.

It can be daunting when you're new to something, so we want to bring you the know-how from the people who live and breathe the outdoors. To help guide us, our team of awe-inspiring OS Champions share their favourite places to visit, hints and tips for outdoor adventures, as well as tried and tested accessible, family- and wheelchair-friendly routes. We hope that you will feel inspired to spend more time outside and reap the physical and mental health benefits that the outdoors has to offer. With our handy guides, paper and digital mapping, and exciting new apps, we can be with you every step of the way.

To find out more visit os.uk/getoutside

RESPECTING
THE COUNTRYSIDE

You can't beat getting outside in the British countryside, but it's vital that we leave no trace when we're enjoying the great outdoors.

Let's make sure that generations to come can enjoy the countryside just as we do.

Leave no trace	Keep dogs under control; bin and bag waste
Do not light fires; only BBQ at official sites	Leave gates as you find them
Keep to footpaths and open access land	Plan ahead for your trip

For more details please visit
gov.uk/countryside-code

USING THIS GUIDE

Easy-to-follow North York Moors walks for all

Before setting off
Check the walk information panel to plan your outing
- Consider using **Public transport** where flagged. If driving, note the satnav postcode for the car park under **Parking**
- The suggested **Time** is based on a gentle pace
- Note the availability of **Cafés**, tearooms and pubs, and **Toilets**

Terrain and hilliness
- **Terrain** indicates the nature of the route surface
- Any rises and falls are noted under **Hilliness**

Walking with your dog?
- This panel states where **Dogs** must be on a lead and how many stiles there are – in case you need to lift your dog
- Keep dogs on leads where there are livestock and between April and August in forest and on grassland where there are ground-nesting birds

A perfectly pocket-sized walking guide
- Handily sized for ease of use on each walk
- When not being read, it fits nicely into a pocket...
- ...so between points, put this book in the pocket of your coat, trousers or day sack and enjoy your stroll in glorious countryside – we've made it pocket-sized for a reason!

Flexibility of route presentation to suit all readers
- **Not comfortable map reading?** Then use the simple-to-follow route profile and accompanying route description and pictures
- **Happy to map read?** New-look walk mapping makes it easier for you to focus on the route and the points of interest along the way
- **Read the insightful Did you know?, Local legend, Stories behind the walk** and **Nature notes** to help you make the most of your day out and to enjoy all that each walk has to offer

OS information about the walk

- Many of the features and symbols shown are taken from Ordnance Survey's celebrated **Explorer** mapping, designed to help people across Great Britain enjoy leisure time spent outside

- National Grid reference for the start point
- Explorer sheet map covering the route

OS information
🚶 NZ 592110
Explorer OL26

The easy-to-use walk map

- **Large-scale** mapping for ultra-clear route finding

- **Numbered points** at key turns along the route that tie in with the route instructions and respective points marked on the profile

- **Pictorial symbols** for intuitive map reading, see Map Symbols on the front cover flap

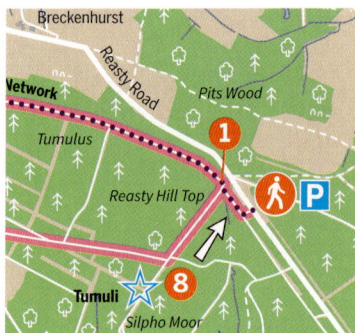

The simple-to-follow walk profile

- Progress easily along the route using the illustrative profile, it has **numbered points** for key turning points and **graduated distance** markers

- Easy-read **route directions** with turn-by-turn detail

- Reassuring **route photographs** for each numbered point

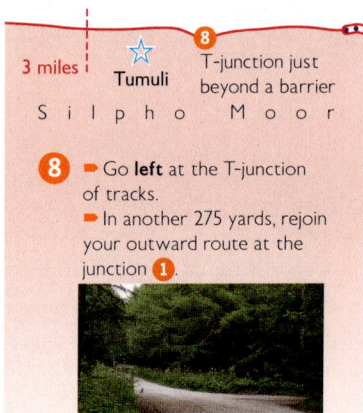

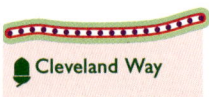

8 ➡ Go **left** at the T-junction of tracks.
➡ In another 275 yards, rejoin your outward route at the junction **1**.

Using QR codes

- Scan each QR code to see the route in Ordnance Survey's OS Maps App.
NB You may need to download a scanning app if you have an older phone

- OS Maps will open the route automatically if you have it installed. If not, the route will open in the web version of OS Maps

- Please click **Start Route** button to begin navigating or **Download Route** to store the route for offline use

WALK 1

COOK'S MONUMENT AND ROSEBRRY TOPPING

GO BY TRAIN

This richly rewarding walk is also the hardest route in the book in terms of the distance and ascent involved, but each climb brings superb views and an uplifting sense of space. It first rises to Captain Cook's Monument, then drops through pine woods before heading uphill again to the base of Roseberry Topping, Yorkshire's own little Matterhorn. You then get to end the day by striding out along wonderful moorland tracks.

OS information

NZ 592110
Explorer OL26

Distance
4.5 miles/7.2km

Time
2¾ hours

Start/Finish
Gribdale Gate

Parking TS9 6HW
Forestry England
Gribdale Gate
roadside parking
area, Dikes Lane,
2 miles east of Great
Ayton

Public toilets
None; nearest in
Great Ayton

Cafés/pubs
Picnic benches
at parking area;
otherwise, Great
Ayton

Terrain
Surfaced paths
and tracks; rough
woodland paths; lane;
steps at the end

Hilliness
Undulating with
several sustained
climbs and descents,
steep in places

Footwear
Year round 👢

Did you know? Easby Moor was the scene of an air crash during World War II. On 11 February 1940, an aircraft took off from Thornaby airfield heading for Denmark to search for German minesweepers. Ice had formed on the wings which made the aeroplane too heavy to gain height causing it to crash into a larch plantation. Three of the four crewmen were killed and the fourth was badly injured.

Local legend A well that once existed on Roseberry Topping was said to have healing properties. If the clothing of a sick person floated in the water, they would recover, and a strip of the cloth would be hung on nearby bushes in thanks. If the clothing sank, then the prognosis was poor. The water was said to cure rickets, lameness and various other ailments, including sore eyes.

Scan Me

Walk 1 Cook's Monument and Roseberry Topping   **15**

STORIES BEHIND THE WALK

☆ **Captain Cook's Monument** The monument to explorer Captain James Cook, who grew up in nearby Great Ayton, was erected on Easby Moor in 1827 by Whitby banker Robert Campion. It commemorates Cook's 'nautical knowledge' and his 'zeal, prudence and energy'. The monument has been restored twice since it was constructed. In 1895, the stonework was repointed and a lightning conductor added. The monument was struck by lightning in 1960, after the conductor had been stolen, and required further restoration.

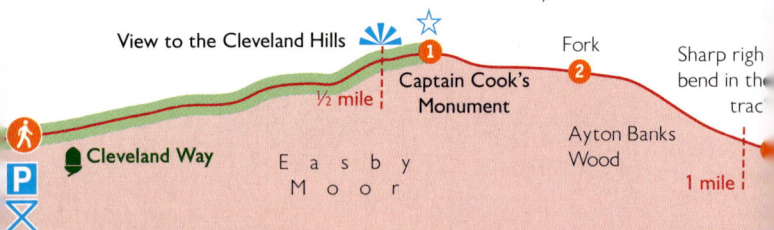

🔱 **Cleveland Hills** The edge of Easby Moor, on which the Captain Cook Monument is located, provides a great view across to the main ridge of the Cleveland Hills. This rises abruptly to form a steep, north-facing escarpment. The highest point is Round Hill on Urra Moor, which is 1,489 feet above sea level. Other notable points in the group include Carlton Bank, Cold Moor and the Wain Stones on Hasty Bank.

View to the Cleveland Hills

1 Captain Cook's Monument

½ mile

Fork

2

Sharp right bend in the track

Cleveland Way

E a s b y
M o o r

Ayton Banks Wood

1 mile

Forestry England
Gribdale car park

▪ Go through the pedestrian gate at the top end of the parking area.
▪ Follow the path uphill for just under ⅔ mile to Captain Cook's Monument.

1 ▪ Reaching the monument, turn sharp **right** – almost back on yourself. The path soon passes between two old stone gateposts.
▪ Keep **forward** for about 150 yards to a fork in the path.

☆ Roseberry Topping

Roseberry Topping can be seen from many points on this walk, but the track from Aireyholme Farm provides the best view of the rocks that give it its distinctive shape. On top of the sandstone base is a layer of ironstone, then a layer of mudstone with a sandstone cap. The surrounding area has similar layering but without the tough sandstone cap, which has left Roseberry Topping standing tall.

☆ Cleveland Way

About half of the walk coincides with the Cleveland Way, a National Trail marked by acorns on signposts. Officially opened in May 1969, the route runs for 109 miles from Helmsley to Filey Brigg and was England's second recognised National Trail. From Helmsley, the path crosses the North York Moors to the coast at Saltburn-by-the-Sea. It then heads south, passing Whitby Abbey, Robin Hood's Bay and the popular seaside town of Scarborough.

5 Aireyholme Farm

2 miles

1½ miles

4 Great Ayton Station, left, 400 yards

2 ➤ At the fork, branch **left**, downhill. The path descends steeply through Ayton Banks Wood.
➤ After emerging from the wood, carry on to a right-hand bend in the track in about 100 yards.

3 ➤ Follow the path round the sharp **right** bend.
➤ When the path ends, continue **ahead** on the surfaced lane to reach a crossroads with Dikes Lane.

NATURE NOTES

Of the various habitats encountered on this walk, the most significant are the forest and the open moorland. The tree species include various conifers, planted for commercial reasons, as well as oak, sycamore, birch, rowan, holly and hawthorn. The paths are lined with plants such as burdock, selfheal, gorse, harebell, bramble, herb robert and ragwort. Listen for goldcrests, great spotted woodpeckers, green woodpeckers and the distinctive song of the increasingly rare yellowhammer, usually rendered as 'a little bit of bread and no cheese'.

On more open ground, particularly along escarpment edges, kestrels hover, waiting to catch their next meal – mice, shrews, small birds, beetles and worms – while the heather, bilberry and moorland grasses hide adders, which come out to bask on sun-kissed paths at the first sign of spring.

Yellowhammer

Roseberry Topping
(summit, left, ¼ mile

6

☆

7

🌲 **Cleveland Way**

Newton Moor (left

3 miles

2½ miles

4 ➨ At the crossroads, go **straight over** onto Aireyholme Lane.
➨ After a sustained climb, the lane bends **left** to pass in front of Aireyholme Farm.
➨ Go through the gate and continue past the farm buildings to a track junction.

5 ➨ At the junction, turn **right** along a rough farm track.
➨ About ⅔ mile beyond Aireyholme Farm, the track goes through a gate, reaching a path junction on open moorland.

Above: adder
Below: gorse

Top: holly
Above: acorns
Below: oak leaves

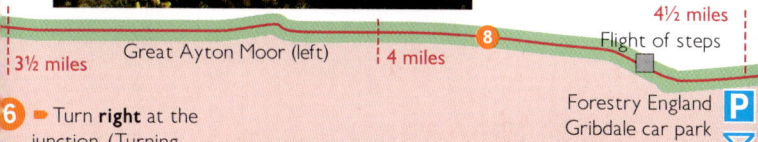

3½ miles Great Ayton Moor (left) 4 miles ⑧ 4½ miles
Flight of steps

Forestry England
Gribdale car park 🅿 ✕

6 ➡ Turn **right** at the junction. (Turning left would take you to the summit of Roseberry Topping – in total, ½ mile detour, climbing another 265 feet.)
➡ The path soon climbs steeply to the next gate, reached in almost ⅓ mile.

7 ➡ Pass through the gate and fork **right** to stay on a clear, easy-going, stony path (following a wall on the right) along the edge of Great Ayton Moor for 1 mile to a three-way fork.

8 ➡ Bear **right** at the fork, keeping by the moorland boundary. The path descends gently at first, but then negotiates a flight of steps.
➡ Drop to the road opposite the parking area.

OSMOTHERLEY

The attractive village of Osmotherley lies on the western edge of the North York Moors National Park. Once the location of a thriving linen industry, the village is more peaceful now, the mills having been replaced by pubs, cafés and B&Bs. Using a wildflower-fringed lane with moorland views, this walk climbs above the honey-coloured cottages to join a section of the Cleveland Way. On the return, the flat expanse of the Vale of Mowbray can be glimpsed through the trees.

OS information

SE 455972
Explorer OL26

Distance
3.8 miles/6.1 km

Time
2¼ hours

Start/Finish
Osmotherley

Parking DL6 3AG
Roadside parking in Osmotherley

Public toilets
Osmotherley, at the junction of South End and School Lane

Cafés/pubs
Osmotherley

Terrain
Pavement; surfaced lane; woodland path; rough tracks

Hilliness
Long, steady ascent and descent

Footwear
Spring/Autumn/Winter
Summer

Public transport
Bus services 80/89, Northallerton to Stokesley, stop at Osmotherley Green at:
hodgsonsbuses.com

Did you know? Nearby Mount Grace Priory, the best-preserved Carthusian monastery in England, is home to the most famous pack of native stoats in Britain. In 1996, an episode of BBC's *Wildlife on One* featured the first ever footage of stoats in their natural environment. Narrated by David Attenborough, the film shows the stoats in the meadows around the site and playing among the ruins.

Local legend The well on Roseberry Topping (see Walk 1) is reputedly the site of the drowning of Oswy – the son of Oswald, King of Northumbria. A prophecy said the boy would drown, so his mother took him to high ground for safety. When she fell asleep, he wandered off, fell into the well and perished. Some legends suggest Osmotherley takes its name from Oswy's mother, who is supposedly buried in the churchyard next to her son.

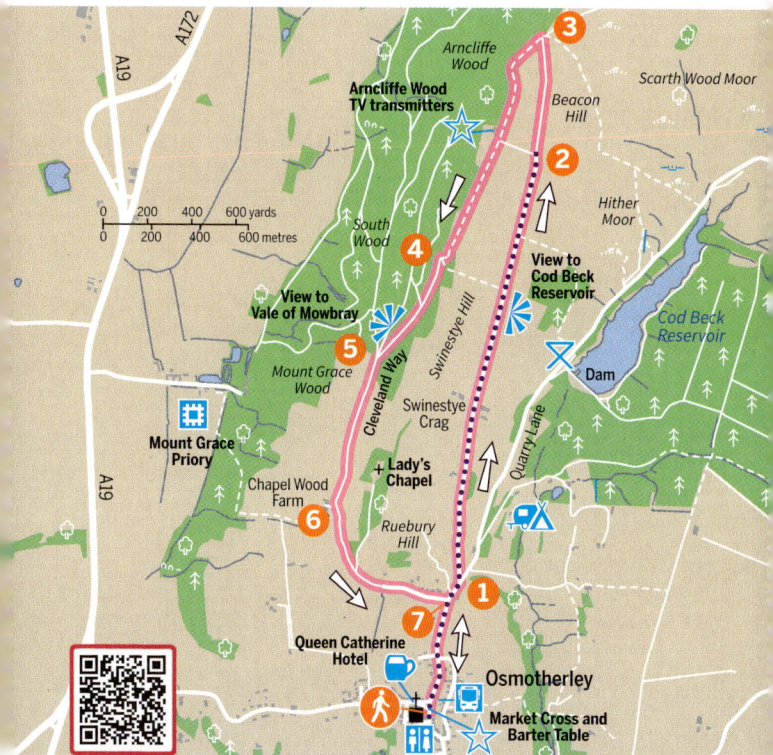

STORIES BEHIND THE WALK

☆ **Market Cross and Barter Table** The original Market Cross in the village dated from the 14th century, but the current one was erected in 1874. Next to it is the 16th-century 'Barter Table'. A slab of sandstone on short legs, it was used as a market stall for the sale of dairy produce until the market ceased operating in the early 19th century. John Wesley, one of the founders of the Methodist movement, preached at the Barter Table in the mid-18th century.

Market Cross and Barter Table

☆ Queen Catherine Hotel

🌿 **Cod Beck Reservoir**
Cod Beck Reservoir was completed in 1953 and a water treatment works was based here until 2006, when poor water quality led to the decision to close it down. It is still owned and maintained by Yorkshire Water and is home to a network of woodland paths. Visitors can walk round the reservoir in under an hour on a purpose-built 1½-mile trail (yorkshirewater.com).

National speed limit sign

½ mile

North End 🌳 **Cleveland Way**

1

🏃 From the village centre, with your back to the Queen Catherine Hotel, head **right** and immediately **left** at the T-junction beside the Market Cross.

🏃 Walk up North End for ⅓ mile to a national speed limit road sign.

1 🏃 At the sign, take the surfaced lane rising **left**. (It has a gate across it.)

🏃 Follow the lane for just over 1 mile to a sharp left-hand bend.

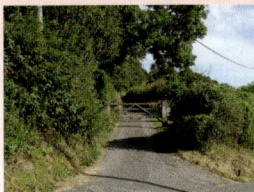

☆ Arncliffe Wood TV transmitters

The compound in Arncliffe Wood is home to an array of masts. Chosen in 1951 for its elevated position, it formed part of a television signal transmission network from Manchester to Kirk O'Shotts, south of Edinburgh. By the 1970s, more than a dozen individual towers had been erected.

☀ Vale of Mowbray

As you walk through Arncliffe Wood, you get glimpses of the Vale of Mowbray through the trees to the right. The floodplain of the River Swale and its tributaries, this flat area is bounded by the high ground of the North York Moors to the east and the hills of the Yorkshire Dales to the west. Its patchwork of fields, divided by hedgerows, is largely given over to arable farming.

View to Cod Beck Reservoir

1 mile

2 Lane bends sharp left · 1½ miles

Kissing-gate and Cleveland Way Fingerpost

3 Arncliffe Wood

2 ► Leave the lane at the bend by turning **right**, through the **leftmost** of two gates.
► After the next gate, continue **straight on** with the wall on your left, rising to a kissing-gate (left) and Cleveland Way fingerpost at the top.

3 ► Go **left** through the kissing-gate and strike out along the Cleveland Way.
► Pass to the **left** of a fenced compound of masts and maintain direction for ⅓ mile to a path fork.

NATURE NOTES

According to the wildlife charity, Froglife, Cod Beck Reservoir is one of Britain's top ten breeding sites for common toads, so there's a chance of spotting them on this walk, particularly when they leave their hibernation sites. Every spring, 'toad patrol' volunteers take to the road leading out of Osmotherley to ensure the toads can cross safely.

The gated lane followed at the start of the walk is lined by trees and hedgerows containing hawthorn, hazel, ash, elder, holly, bramble and rowan. A splash of summer colour along path and road verges is provided by red clover, the purple of tufted vetch, the blue of harebells, the yellow of gorse, the bright pink spikes of rosebay willowherb and the deep pink of foxglove. Although the latter is poisonous if consumed directly, it also contains digitalis, a chemical used to treat heart conditions.

The woodland fringes and hedgerows are a favoured habitat for long-tailed tits. Their tales are almost as long as their bodies.

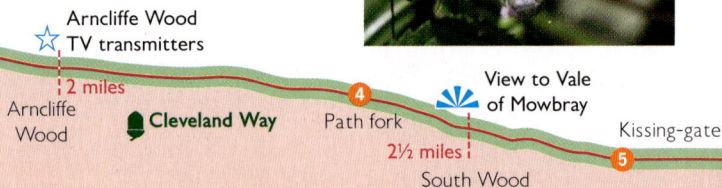

Arncliffe Wood
TV transmitters

2 miles

Arncliffe Wood

Cleveland Way

Path fork

4

View to Vale of Mowbray

2½ miles

South Wood

Kissing-gate

5

4 ➤ Bear **right** at the fork. The route now begins descending.
➤ Carry on along the Cleveland Way to break out of the woodland in ⅓ mile, reaching a kissing-gate just after a track merges from the right.

5 ➤ Pass through the gate and keep **straight on** through the first field.
➤ The path becomes more obvious in the next field. Continue through several more gates to reach a track junction near Chapel Wood Farm.

Top left: common toad
Above: red clover
Top right: foxglove
Opposite page:
bramble flowers

Rosebay willowherb

3 miles
6
Track junction at
Chapel Wood Farm

Lane
3½ miles
7
North End
Queen Catherine Hotel

Market Cross
and Barter Table

6 ➤ Keep **left** along the track.
➤ After a little over ⅓ mile,
the track drops to the road
(North End), followed earlier
out of the village.

7 ➤ Turn **right** to return to the
village centre.

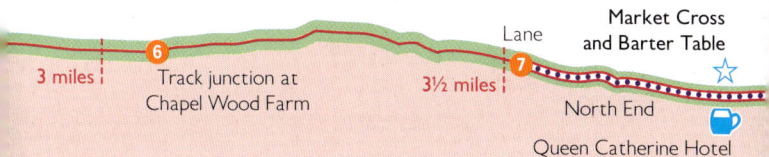

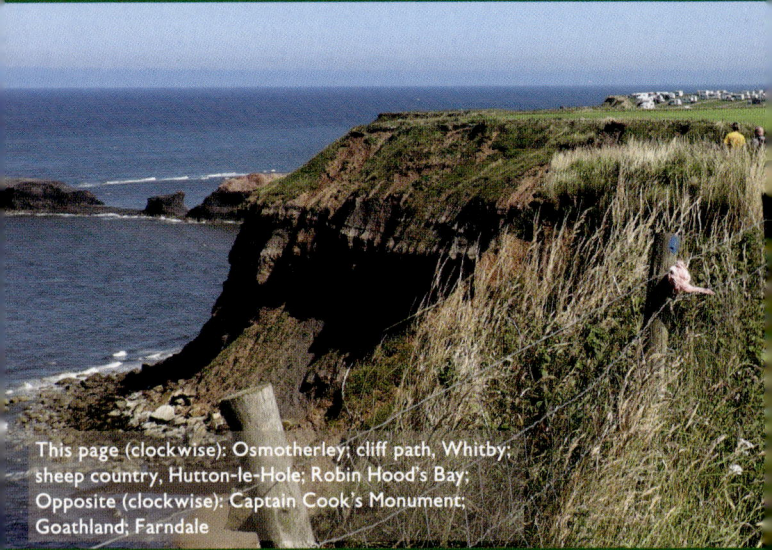

This page (clockwise): Osmotherley; cliff path, Whitby; sheep country, Hutton-le-Hole; Robin Hood's Bay; Opposite (clockwise): Captain Cook's Monument; Goathland; Farndale

SLOW DOWN

WALK 3

SUTTON BANK AND KILBURN WHITE HORSE

The view from Sutton Bank was described by *All Creatures Great and Small* author James Herriot as the 'finest in England'. Was he right? Decide for yourself by walking, from the National Park Centre, along the excellent path skirting the escarpment edge. After passing over the top of the Kilburn White Horse, the walk returns via a series of woodland paths and quiet lanes. Watch for gliders taking off and soaring in the skies above.

OS information

⊕ SE 516830
Explorer OL26

Distance
3.6 miles/5.8km

Time
2 hours

Start/Finish
Sutton Bank
National Park Centre

Parking YO7 2EH
National Park car
park (charge)

Public toilets
In the National Park
Centre

Cafés/pubs
Picnic benches at
⊕; Park Life Café
at National Park
Centre; Gliders' Nest
Café at Yorkshire
Gliding Club

Terrain
Surfaced paths; quiet
lane; woodland trail

Hilliness
Almost flat

Footwear
Winter 🥾
Spring/Autumn/
Summer 👟

🚌 **Public
transport**
Limited bus service,
M4, Redcar to
Sutton Bank via
Stokesley, runs on
summer Saturdays
(one bus in each
direction): moorsbus.
org

Did you know? Kilburn White Horse is the largest white horse in Britain by surface area. It is 220 feet tall and 318 feet long and is said to be visible from the northern side of Leeds, about 30 miles away. During World War II, it was covered over as it was thought that German bomber crews would be able to use it as a navigational aid.

Local legend A spectral figure in a black dress haunts Sutton Bank. She has been seen lying beside the road and also flagging down traffic. She may be the ghost of Abigail Claister, who lived in Kilburn in the early 17th century and was believed to be a witch. Pursued by locals, she ran up Sutton Bank flagging down passersby, but nobody helped her. In desperation, she leapt to her death from Whitestone Cliff.

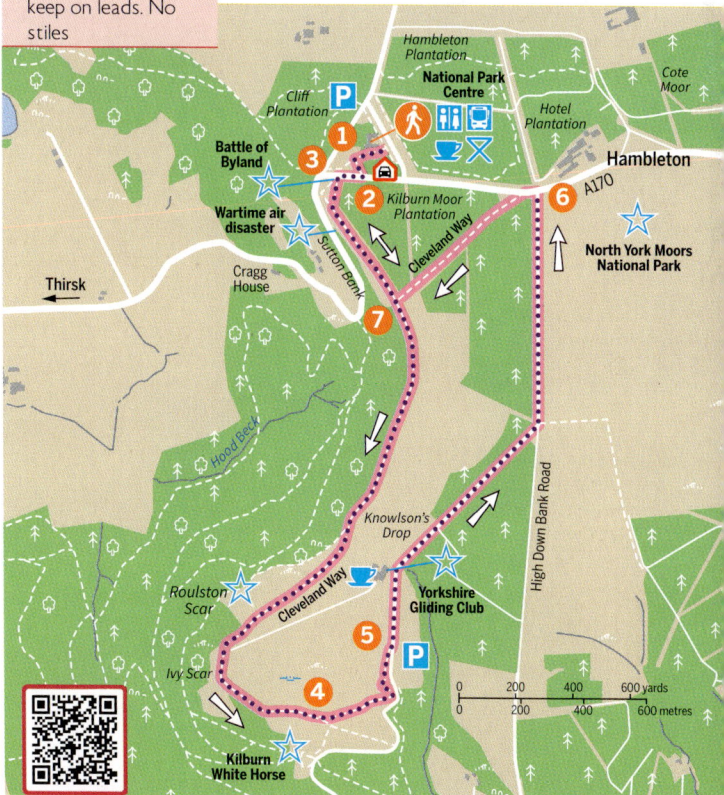

STORIES BEHIND THE WALK

☆ **Battle of Byland** Sutton Bank was the site of the Battle of Byland in 1332, part of the First Scottish War of Independence. Scottish forces commanded by Robert the Bruce had crossed into England and marched to Northallerton. Hoping to catch Edward II's depleted forces unawares, they raced to Sutton Bank. The English at first used the height of the escarpment to their advantage but, after fierce fighting, the Scots claimed victory.

North York Moors National Park

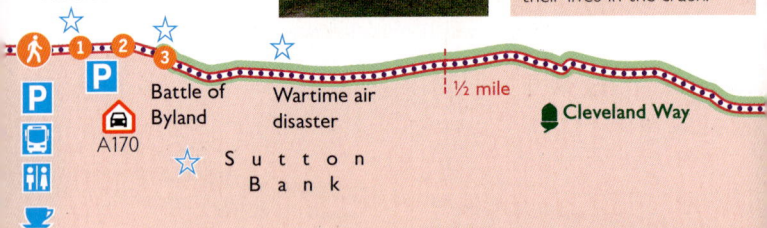

☆ **Wartime air disaster**

A memorial beside the path commemorates air crews who perished in World War II, including the crew of Halifax bomber JD 105. In the early hours of 5 May 1943, the plane was part of a squadron returning from a flight to Dortmund, when they were diverted to Leeming airfield. In poor visibility, the pilot clipped the top of Hood Hill causing the aircraft to split apart. Five of the eight-member crew lost their lives in the crash.

Battle of Byland

Wartime air disaster

½ mile

Cleveland Way

S u t t o n B a n k

A170

Sutton Bank National Park Centre car park

➡ With your back to the National Park Centre, go **right** and turn **right** again — round the side of it — to a path junction just left of the far end of the building.

1 ➡ At the path junction, go **left** and then **left** again just before a car park — signposted Cleveland Way and White Horse Trail — to the main road (A170).

☆ **North York Moors National Park** The North York Moors was designated as a National Park in 1952. Covering an area of 554 square miles, it contains one of the largest expanses of heather moorland in Britain, as well as woodland, impressive sandstone escarpments and 26 miles of rugged coast. It has a total of 1,408 miles of public rights of way and receives about 12.3 million day visits per year (northyorkmoors.org.uk).

☆ **Yorkshire Gliding Club** Sutton Bank has been used as a glider launching site since 1911, although the Yorkshire Gliding Club wasn't established until 1934. The airfield is 900 feet above sea level and 600 feet above the bottom of the cliffs. The site benefits from a rising air flow due to wind being forced upwards as it hits the escarpment. Gliders were initially launched by men hauling them with ropes, but the club now has an engine and winch (ygc.co.uk).

1 mile   Roulston Scar   1½ miles ☆   Kilburn White Horse

2 ► Carefully **cross** the main road and follow the path to the **right** for about 75 yards to a **left-hand** bend.

3 ► Immediately round the bend the route ahead forks. Bear **left**.
► The surfaced path leads on to skirt the edge of the cliffs for 1⅓ miles, reaching a fork directly above the Kilburn White Horse.

NATURE NOTES

The early part of the walk passes through birch woodland and heathland along the top of high cliffs. You are likely to see jackdaws on the crags along Sutton Bank, as well as buzzards, the occasional raven and, if you're lucky, red kites. The birch, oak, sycamore, hawthorn, gorse and lichen host a variety of invertebrates, which are food for birds such as marsh tits, whitethroats, robins, blue tits, siskins and yellowhammers. Turtle doves have also been seen in this area after returning from overwintering in West Africa. Numbers of this once relatively common bird have begun to recover throughout Europe following a hunting ban that protects them.

The sides of the path and road verges are rich in wildflowers, shrubs and trees, including spear thistle, bird's-foot trefoil, red campion, selfheal, marsh woundwort, meadow crane's-bill, harebell, hedge bindweed, hazel, alder and bilberry. There are also patches of Himalayan balsam, brought into Britain as an ornamental plant and now aggressively displacing native species.

☆ Yorkshire Gliding Club

2½ miles

2 miles

4 🔹 Bear **left** at the path fork above the white horse.
🔹 This path later bends **left** and winds through light woodland to reach a lane.

5 🔹 Turn **left** along the quiet lane, noting the Yorkshire Gliding Club (left) in 300 yards.
🔹 After 1 mile of road walking, you reach a T-junction with the A170.

Jackdaw

Top left: siskin
Top middle: whitethroat
Top right: harebells
Opposite page: turtle dove

Cleveland Way fingerpost

A170

3 miles

T-junction

Wartime air disaster

S u t t o n
B a n k

Battle of Byland

North York Moors National Park

3½ miles

A170

Sutton Bank National Park Centre car park

6 ➤ Don't cross the road; instead, take the Cleveland Way signposted to the **left**.
➤ The grassy trail descends a short embankment and becomes clearer, running on for ⅓ mile to a Cleveland Way T-junction.

7 ➤ At the T-junction (with the surfaced path followed earlier in the walk), turn **right**.
➤ Retrace your steps to the National Park Centre, ½ mile away.

RIEVAULX ABBEY

The tiny settlement of Rievaulx, with its romantic abbey ruin, is tucked away in Ryedale to the north-west of Helmsley. This gentle walk explores a small section of the peaceful, wooded dale. After briefly heading upstream from the village and crossing Bow Bridge, it enters the atmospheric Ashberry Wood. The going gets rougher underfoot here, but the outing ends with a stroll along quiet lanes and past the abbey, which has a museum and café.

OS information	
🏃 SE 574850 Explorer OL26	
Distance	
2.6 miles / 4.1km	
Time	
1½ hours	
Start/Finish	
Rievaulx	
Parking YO62 5LB	
Village car park (honesty box), 100 yards north of the abbey car park	
Public toilets	
In the abbey car park	
Cafés/pubs	
Rievaulx Abbey Café; Helmsley, 2½ miles to the south-east	
Terrain	
Lanes; tracks; woodland path, rough in places	
Hilliness	
Undulating; steady ascent/ descent 🏃 to ❸; a steeper rise/fall in Ashberry Wood ❹ to ❻	
Footwear	
Winter/Spring/ Autumn 👢 Summer 👟	
🚌 **Public transport**	
Limited bus service, M4, Redcar to Sutton Bank via Stokesley, stops at the abbey on summer Saturdays (one bus in each direction): moorsbus.org	

Did you know? Rievaulx Abbey was once home to an ironworks. The 'Yron Smiths' was a water-powered forge within the abbey precincts that would have produced nails and other iron implements for use in the abbey. After Henry VIII's Dissolution of the Monasteries, the ironworks grew, and a blast furnace was added in the late 16th century. The operation finally closed in the mid-17th century due to lack of local timber.

Local legend Three miles east of Rievaulx lies the impressive Helmsley Castle. The 12th-century ruin is said to be haunted by two ghosts. One is a lady in a green dress who has been seen inside and outside the main castle buildings. The other is supposedly the ghost of a Civil War soldier who died of starvation during a siege at the castle.

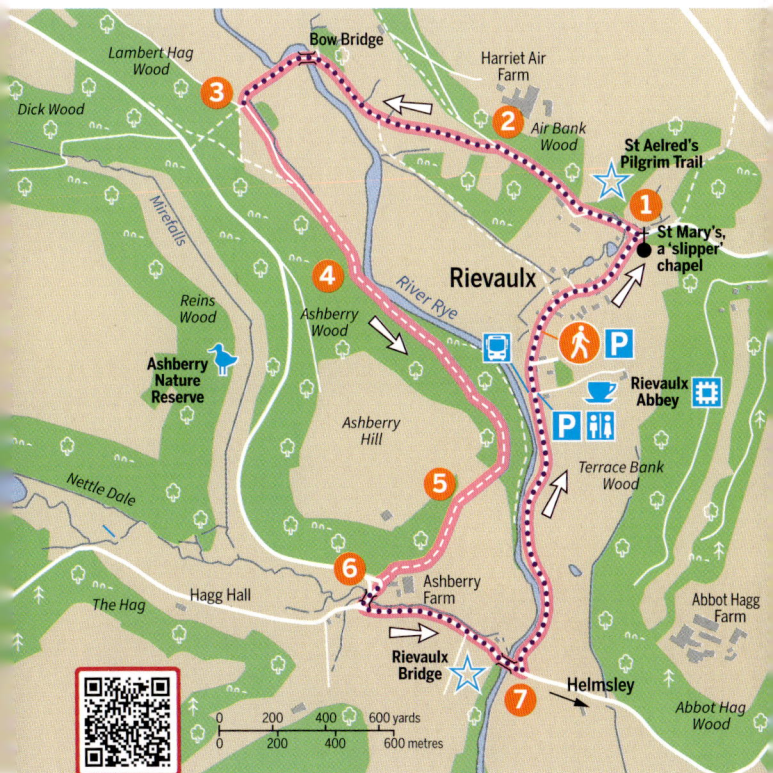

Scan Me

STORIES BEHIND THE WALK

✝ St Mary's, a 'slipper' chapel

St Mary's parish church was once the 'slipper' chapel for Rievaulx Abbey – a popular destination for pilgrims who visited the shrine of St William, the abbey's first abbot. This small chapel, originally built into the abbey gates, was where pilgrims would remove their shoes or boots and put on slippers before proceeding to the main part of the abbey.

☆ St Aelred's Pilgrim Trail

This walk follows parts of the St Aelred's Pilgrim Trail, a 41-mile route linking all the churches in the Helmsley and Upper Ryedale benefice. Named after Aelred, who was abbot of Rievaulx Abbey from 1147 to 1167, it begins and ends in Helmsley. On the way, it visits Sproxton, Scawton, Old Byland and Bilsdale, with Rievaulx forming part of a loop on the trail.

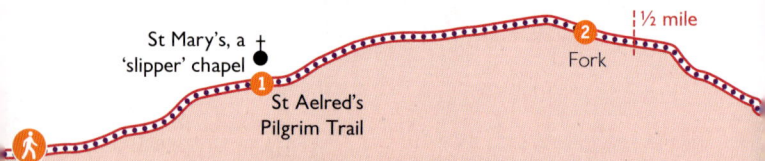

St Mary's, a ✝
'slipper' chapel ●

1 St Aelred's
Pilgrim Trail

2 Fork

¦ ½ mile

🅿
Village
car park

● Leave the car park by turning **right** along the lane.
● Continue for 350 yards, uphill through the village to a junction just beyond the church.

1 ● At the junction, take the rough lane on the **left**.
● As the lane climbs slightly, keep **right** as a private driveway goes left, and carry on to the next fork in about 300 yards.

☆ **Rievaulx Bridge** A three-arch limestone structure, Rievaulx Bridge was built in 1754 to replace a medieval bridge that was washed away by flooding. JMW Turner stood on this bridge to capture a view of the abbey – the picture is now on display in York Art Gallery. During World War II, a concrete ramp was installed allowing tanks from the military camp at Duncombe Park to ford the river without damaging the bridge.

🔲 **Rievaulx Abbey**
Founded in 1132, Rievaulx Abbey was the first Cistercian abbey in the north of England. Its most famous abbot was Aelred, who became a well-known writer and scholar. Under Aelred, the population of the abbey rose to 650 – 150 monks and around 500 lay brothers. The abbey was closed in 1538 as part of Henry VIII's Dissolution and was sold to the Earl of Rutland who dismantled many of the buildings.

Bow Bridge; River Rye   Gate   **3**   Gate into woodland **4**

1 mile

2 🢒 At the fork bear **left**, heading downhill.
🢒 In ¼ mile, the route crosses Bow Bridge.
🢒 200 yards beyond the bridge, look for a gate on the left as the track bends right.

3 🢒 Turn **left** through the gate.
🢒 After another gate, walk along the meadow's right-hand edge, rising to a gate into woodland.

NATURE NOTES

Ashberry Wood is an area of ancient, semi-natural woodland where tree species include sycamore, beech, holly, hawthorn, ash, elder, horse chestnut and crab apple. In spring, the woodland floor is carpeted with wildflowers, including primrose, wood anemone, bluebell and wild garlic. These all flower before the leaves of the woodland canopy block out most of the sunlight. Later in the summer, you should be able to find spear thistle, burdock, meadow crane's-bill, selfheal and meadowsweet on the woodland edges. Both fallow and roe deer might be seen among the trees and grazing in the adjoining meadows.

The age of the woodland here gives rise to a good range of mosses. Decaying branches are also home to the grey spikes of candlesnuff fungus for most of the year; while in late summer and autumn other mushrooms and toadstools such as puffballs and shaggy inkcaps may be encountered.

Top: puffball
Above: shaggy inkcap

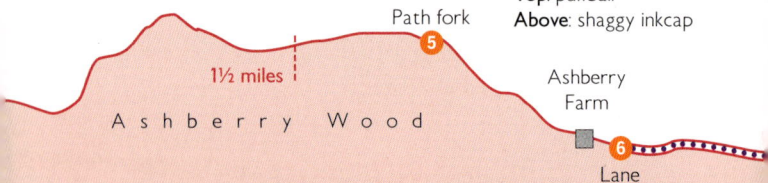

Path fork

5

1½ miles

A s h b e r r y   W o o d

Ashberry
Farm

6

Lane

5 ➤ Keep **left** at the fork, along the lower path, as the other path climbs right.
➤ Turn **right** after the gate leading on to Ashberry Farm's driveway and walk out to a lane.

4 ➤ Pass through the gate and follow the path through the trees for almost ½ mile to a fork.

Above: candlesnuff fungus **Below:** roe deer

Above: meadowsweet
Below: beech leaves

Rievaulx
Abbey
2½ miles

⭐ **Rievaulx Bridge**
River Rye

miles

6 ➡ Go **left** along the lane, soon crossing a small bridge.
➡ At the lane junction just beyond, turn **left** – signposted Rievaulx – to the next junction, immediately beyond Rievaulx Bridge.

Village car park

7 ➡ Go **left** and return along the lane. The car park is on the right in ½ mile.

Walk 4 Rievaulx Abbey   **39**

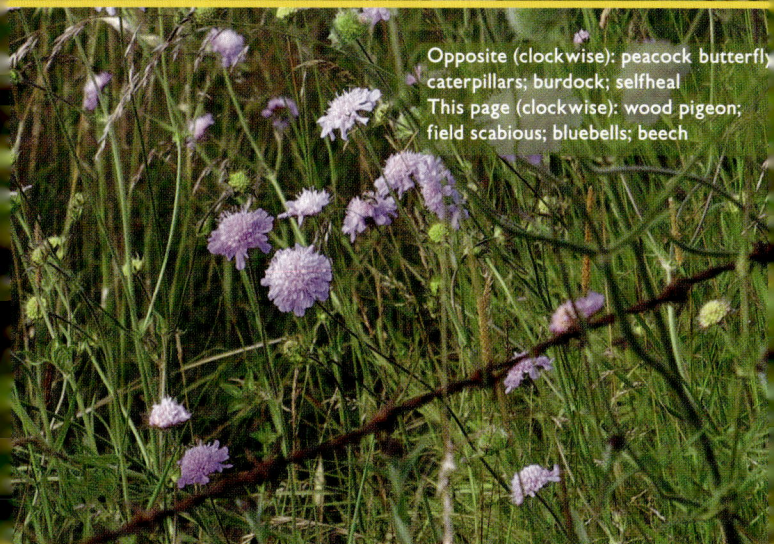

Opposite (clockwise): peacock butterfly caterpillars; burdock; selfheal
This page (clockwise): wood pigeon; field scabious; bluebells; beech

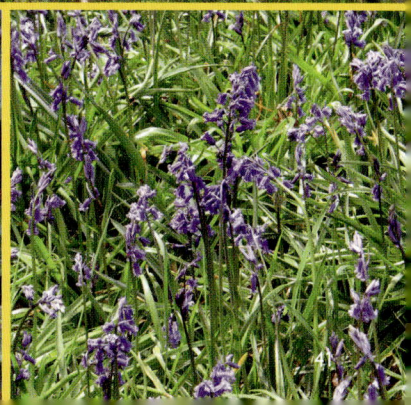

WALK 5

FARNDALE

The first part of this walk, beside the meandering River Dove, is known as the Daffodil Walk. In early spring, the dale comes to dazzling life as wild daffodils burst into bloom. But the walk is delightful all year round with the river, woodland and meadows in serene harmony – and all enjoyed from an easy-going surfaced path. After visiting High Mill, return via the sheep-filled pastures that climb the valley sides for a fresh perspective on this lovely dale.

OS information

🚶 SE 672952
Explorer OL26

Distance
2.9 miles/4.6km

Time
1¾ hours

Start/Finish
Low Mill, Farndale

Parking YO62 7UY
Car park beside The Band Room (honesty box when no staff on duty)

Public toilets
In the car park

Cafés/pubs
Daffy Caffy, High Mill, at **1**; Farndale Store, at 🚶, serves hot drinks and snacks to take away

Terrain
Surfaced path 🚶 to **2**; field paths from **2**, often unclear on the ground but well waymarked

Hilliness
Gently undulating 🚶 to **2**; moderately steep climb **2** to **5** and then gradual descent back to 🚶

Footwear
Winter/Spring/Autumn 🥾
Summer 👟

Did you know? In 1944, a Farndale resident was awarded the Edward Medal. Harwood Henry Flintoff was only 13 when he rescued a farmer who'd been knocked to the ground by a bull. Flintoff wrestled with the bull, and he and the farmer managed to subdue it until further help arrived. In 1971, holders of the Edward Medal were invited to exchange it for the George Cross – the highest award for civilian gallantry.

Local legend A local farmer and his wife awoke one night to the sound of corn being threshed. Discovering a hob – a friendly household spirit – working hard, they rewarded him with their best cream, and he continued working every night. When the farmer's wife died, his new bride only fed the hob skimmed milk. He left in disgust, the farm became cursed, and no labourers would stay. Eventually, the farmer and his wife left the valley.

Public transport
None

Accessibility
Powered wheelchair and all-terrain pushchair friendly to 2, with return via the outward route

Dogs
Welcome but on leads (livestock). No stiles

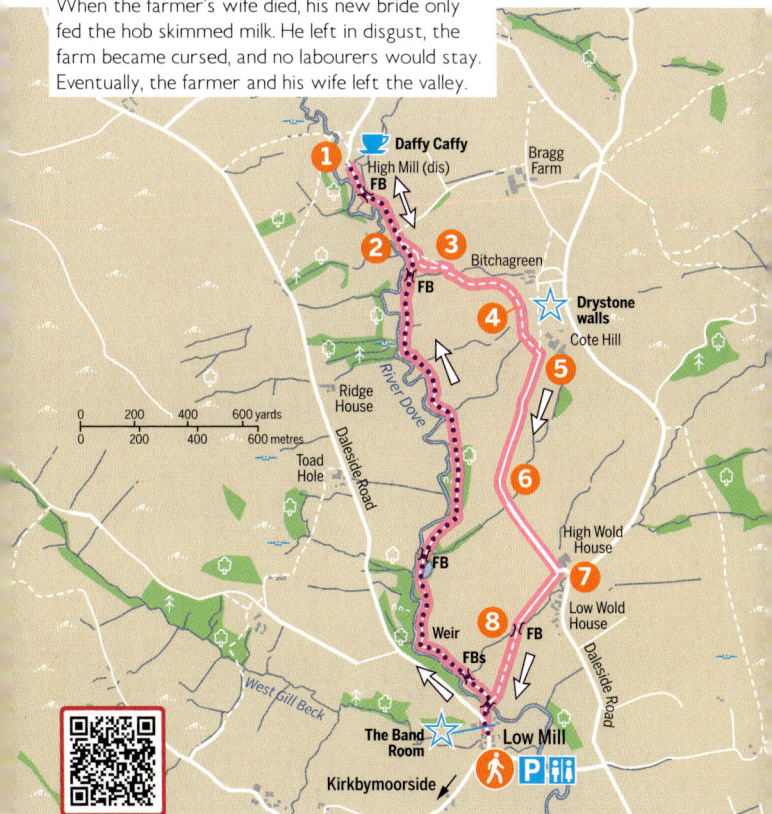

Daffy Caffy
High Mill (dis)
FB
Bragg Farm
Bitchagreen
Drystone walls
Cote Hill
River Dove
Ridge House
Dalesside Road
Toad Hole
High Wold House
Low Wold House
Dalesside Road
Weir
FBs
FB
West Gill Beck
The Band Room
Low Mill
Kirkbymoorside

0 200 400 600 yards
0 200 400 600 metres

STORIES BEHIND
THE WALK

☆ **The Band Room** Originally built for a brass band in the 1920s, The Band Room now regularly hosts musical acts from around the world. The tiny venue – with a maximum capacity of 100 people – mainly hosts contemporary Americana bands but, in 2007, Welsh singer-songwriter Cerys Matthews chose The Band Room to preview her tour. There is no bar, so audience members are encouraged to bring their own refreshments to enjoy.

☆ **Sheep farming**
You're likely to see a lot of sheep in the enclosures on the second half of this walk, including young lambs at daffodil time. Ewes are mated with rams during October and November, so that lambs are born in late winter and early spring. Even on upland farms, many lambs are born outside, although some may be given shelter for the first few days of their lives.

☆ **The Band Room**
Bridge over River Dove

River Dove

½ mile

Low Mill

🅿 🚻

➤ Turn **right** out of the car park and immediately **right** again.

➤ **Cross** a small bridge and then turn **left** through a gate.

➤ A surfaced path heads upstream beside the River Dove for 1⅓ miles to High Mill.

① ➤ Pass through the gate at High Mill. Keep **forward** between the buildings for 50 yards to the Daffy Caffy (right).

➤ From here, retrace your steps through two gates, pausing ahead of the third.

☆ Failed reservoir bid

Farndale would have looked very different if plans to construct a reservoir had come to fruition. The Kingston upon Hull Corporation gained permission to create a huge artificial lake to secure a water supply for the city. Plans stalled during World War II and were later complicated by the creation of the North York Moors National Park. But it wasn't until 1970, when the Bill to create the dam was thrown out of Parliament, that Farndale was saved.

☆ Drystone walls

Farndale's valley sides are divided into a patchwork of fields separated by drystone walls, a common feature of the countryside throughout northern England. Some date from medieval times, although the long, straight walls typical of the higher ground are more closely associated with the enclosure acts of the 18th and 19th centuries. Note how the walls are built without any mortar; stability is achieved through traditional construction methods.

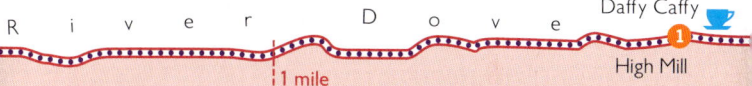

River Dove

Daffy Caffy

High Mill

1 mile

2 ➡ Nearing the third gate, leave the path by heading slightly **left**, through a gate just above the one on the main path.
➡ Walk beside the field boundary (right) for 200 yards, following it to the **left** up to a gate (right).

3 ➡ Turn **right**, through the gate. Walk with a wall on your right, following it round to the **left**.
➡ Go through the gate at the top; bear **right**, walking parallel with the field boundary (right) to a gate between two oaks.

NATURE NOTES

Farndale is famous for its daffodils which, in spring, cover the banks of the River Dove and some of the surrounding meadows. Daffodils were first brought to Britain by the Romans. They grow from bulbs and those in Farndale are thought to have been planted many centuries ago by the monks of Rievaulx Abbey.

Among the many other plants and tree species lining the riverside path are hazel, oak, rowan, marsh woundwort and hart's-tongue fern.

Watch and listen for birds such as long-tailed tits, wrens and robins.

The fields are grazed largely by Swaledale sheep, one of England's most common hill breeds. Characterised by its black face, white nose and curly horns, the Swaledale is a hardy sheep suited to harsh upland environments.

In spring, the distinctive calls of curlews and lapwings can be heard as they protect their ground nests on the valley's higher slopes.

Left: wren
Middle: robin
Right: curlew

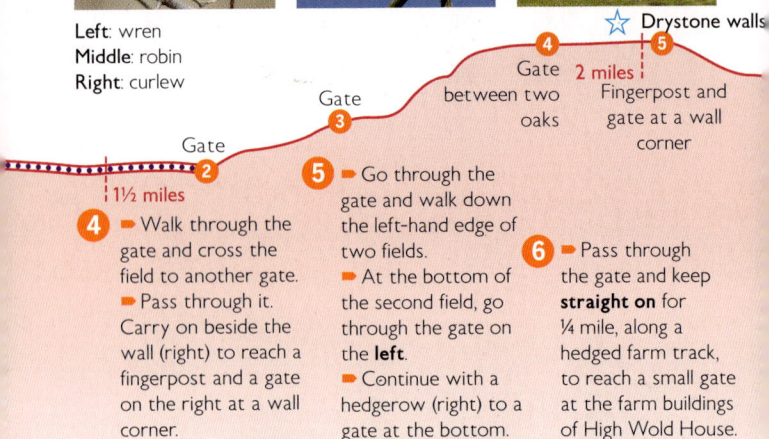

☆ Drystone walls

4
Gate
between two
oaks

2 miles

5
Fingerpost and
gate at a wall
corner

Gate
3

Gate
2

1½ miles

4 ➤ Walk through the gate and cross the field to another gate.
➤ Pass through it. Carry on beside the wall (right) to reach a fingerpost and a gate on the right at a wall corner.

5 ➤ Go through the gate and walk down the left-hand edge of two fields.
➤ At the bottom of the second field, go through the gate on the **left**.
➤ Continue with a hedgerow (right) to a gate at the bottom.

6 ➤ Pass through the gate and keep **straight on** for ¼ mile, along a hedged farm track, to reach a small gate at the farm buildings of High Wold House.

Farndale daffodils

Hart's-tongue fern

Above: marsh woundwort
Below: Swaledale sheep

Gate **6**

Gate at farm buildings, High Wold House **7**

2½ miles

Gate **8**

The Band Room ☆

Bridge over River Dove

🅿

Low Mill

7 ➤ Turn **right** through the gate.
➤ Walk downhill with a wall (right) for 250 yards to a pair of gates.

8 ➤ At the bottom of the enclosure, go through the smaller, left-hand gate.
➤ Maintain a straight line, through trees and across a field, to return to the bridge used earlier.
➤ **Re-cross** the bridge and return to the start.

HUTTON-LE-HOLE

There's lots of variety on this moderately undulating walk starting from the idyllic village of Hutton-le-Hole. It passes through fields, woodland and along the edge of open moorland to reach lovely Lastingham, with its fascinating church. From here, the route heads along quiet lanes to peaceful Spaunton. Having reached the highest point on the walk, with good views of the surrounding moors, the route concludes along wildflower-fringed tracks across arable fields.

OS information

SE 705902
Explorer OL26

Distance
4.3 miles/7km

Time
2½ hours

Start/Finish
Hutton-le-Hole

Parking YO62 6UB
National Park pay-and-display car park

Public toilets
In the car park

Cafés/pubs
Hutton-le-Hole;
Blacksmiths Arms in Lastingham

Terrain
Field and woodland paths; quiet roads; moorland and farm tracks

Hilliness
Moderately undulating; one steeper ascent **5** to **6**, and steep descent **8** to **9**

Did you know? It's hardly surprising that the attractive, well-kept village of Hutton-le-Hole has several listed buildings – 29 in total. However, not all of these are buildings in the traditional sense as they include: the telephone box near the church (a K6 model designed by Sir Giles Gilbert Scott for George V's Silver Jubilee in 1935); a set of railings and its gate; and a sundial.

Local legend There is a standing stone to the north of the village known as Hang Man's Stone. According to local legend, a thief stole and killed a sheep, tying its legs together so he could carry it around his neck. Stopping for a rest, he placed the carcass on a rock – the legs still around his neck. After he fell asleep, the sheep's body slipped, trapping him against the stone and the unlucky thief was strangled.

Footwear
Winter/Spring/
Autumn
Summer

Public transport
Limited bus service M3, Guisborough to Pickering, summer Saturdays only (two buses in each direction): moorsbus.org

Accessibility
Village pavement and lanes only

Dogs
Welcome but keep on leads (livestock). No stiles

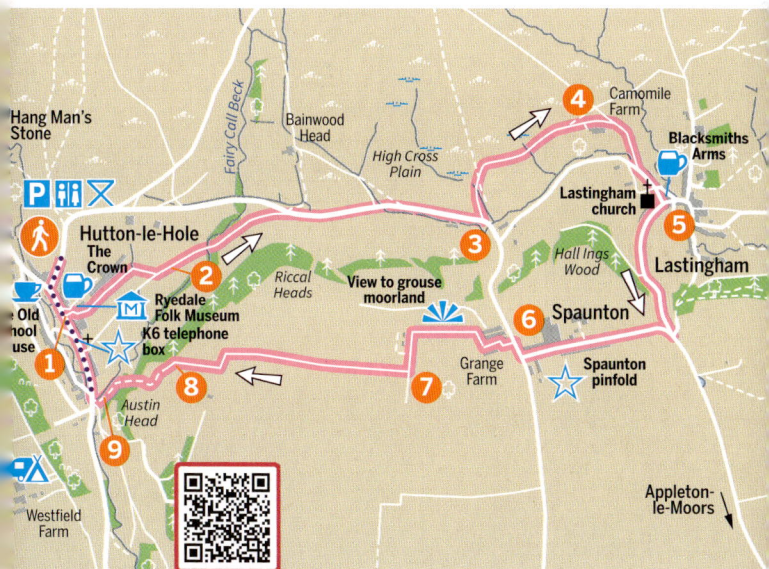

STORIES BEHIND THE WALK

🏛 **Ryedale Folk Museum** Local historian Wilfred Crossland wanted to preserve and share the stories of ordinary Ryedale people. His original collection of objects was in his family home, which Wilfred and his siblings opened to the public. This is now part of the Ryedale Folk Museum. Visitors to the open-air

museum can also see 20 heritage buildings, including a thatched manor house, a vintage chemist, village store and a blacksmith's workshop.

☆ **Grouse moorland** After leaving Spaunton on the farm track, walkers are treated to good views north towards grouse moorland. Red grouse numbers are maintained for game shooting through a habitat-management programme of heather burning, bracken clearing and predator control.

The Old School House

Kissing-gate

½ mile

1 mile

Ryedale Folk Museum

The Crown

P

National Park car park

➤ Leaving the car park, turn **left** along the road.
➤ Go **left** again at the T-junction and continue to the Ryedale Folk Museum (left).

1 ➤ Immediately after the museum, take the track **left** and, beyond a gate, keep **straight on**.
➤ After a kissing-gate and fenced section, skirt the left-hand edge of two fields.
➤ Drawing level with a gate, bear slightly **left**, walking right of hawthorns to a kissing-gate.

2 ➤ Go through the kissing-gate and continue beside a fence on the left.
➤ The next kissing-gate leads into woodland and over a bridge.
➤ Beyond the trees, the path leads to a road. Turn **right** along it for ½ mile to a track (left).

☆ Spaunton pinfold

Like many other villages, Spaunton had a pinfold. It would've been used to impound livestock that had wandered onto arable farmland. The animals were released only when their owners paid a fine to the pinder, whose job it was to round up and feed stray animals. This was a salaried job, sometimes coming with a house and garden. Spaunton's pinfold was restored in 2012.

✝ Lastingham church

St Mary's Church in Lastingham has a feature that is unique in England. Beneath the red-tiled floor lies a Norman crypt, the only one in the country with a nave, apse and aisles. William the Conqueror granted permission for the church in 1078 and it was built on the supposed burial place of Cedd, abbot of Lastingham monastery. The crypt now contains a range of historical artefacts, including parts of a Saxon cross.

3 Rough track   1½ miles   **4** Track   Lastingham church ✝ **5** 2 miles   Blacksmiths Arms

3
- Turn **left** along the rough track. Beyond a gate, the track skirts open moorland.
- Where another path goes left, stay close to the fence on the right.
- When the fence ends, keep **straight on**, dropping to a track.

4
- Bear **left** along the track.
- After a gate, drop to the road and turn **left** along it, entering Lastingham.
- Continue to a junction, with the church (right) on the corner.

NATURE NOTES

Pollinators such as bees, including buff-tailed, white-tailed, red-tailed and common carder bumble bees, as well as butterflies and other insects love the field margins walked on this route. They are most likely to be seen – in spring and summer – on the colourful flowering plants that grow alongside the tracks between **6** and **8**; plants such as chicory, poppies, field scabious, meadow crane's-bill, bramble and hawthorn.

After leaving the woods between **2** and **3**, the route passes through a short section of high bracken. One of several ferns native to Britain, this is thought to be the most common plant in the world. The young fronds, containing hydrogen cyanide, are poisonous to grazing animals.

Spaunton pinfold ☆

Cattle grid **6**

View to grouse moorland

Grassy track **7**

2½ miles

3 miles

5 ➡ Turn **right**, signed Appleton-le-Moors.
➡ Having climbed for ⅓ mile, turn **right** at the next road junction.
➡ Pass through Spaunton and, just after the road bends **right**, look for a cattle grid (left).

6 ➡ Turn **left** and cross the cattle grid. Bear **right** when the concrete track forks.
➡ Beyond the buildings, turn **right** along a rougher track.
➡ After ¼ mile, pass to the **left** of farm sheds and, soon after, reach a grassy track (right).

7 ➡ Turn **right** along the grassy track between fields. This bends **left** after ½ mile.
➡ 75 yards after the bend, turn **right**, leaving the field at a footpath sign.
➡ After 200 yards, the path bends **left**, and 40 yards further, reaches a gap in the trees.

Above: red grouse

Above: buff-tailed bumble bee
Right: red-tailed bumble bee
Far right: chicory
Opposite page: poppy

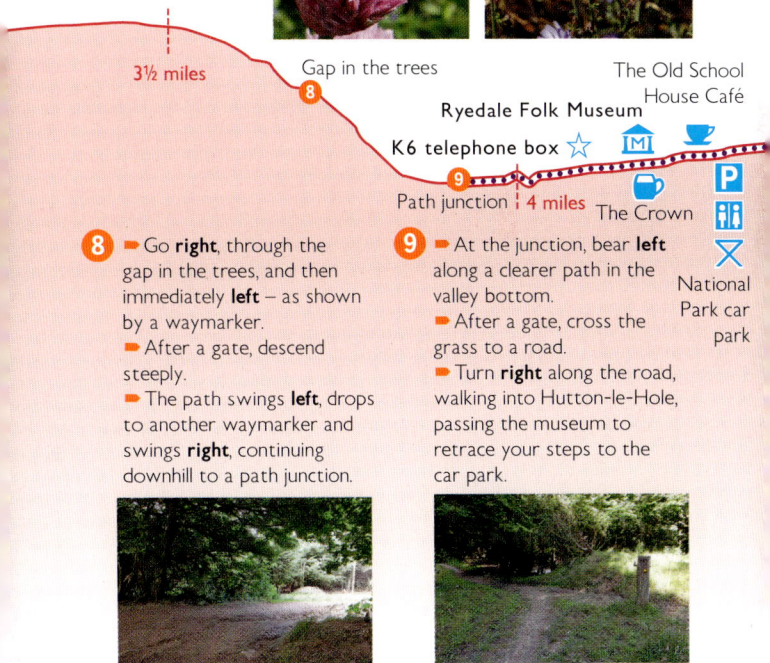

3½ miles

Gap in the trees

The Old School
House Café

Ryedale Folk Museum

K6 telephone box

8

Path junction   4 miles   The Crown

9

National
Park car
park

8 ▶ Go **right**, through the gap in the trees, and then immediately **left** – as shown by a waymarker.
▶ After a gate, descend steeply.
▶ The path swings **left**, drops to another waymarker and swings **right**, continuing downhill to a path junction.

9 ▶ At the junction, bear **left** along a clearer path in the valley bottom.
▶ After a gate, cross the grass to a road.
▶ Turn **right** along the road, walking into Hutton-le-Hole, passing the museum to retrace your steps to the car park.

GROSMONT TO GOATHLAND

This linear walk, making use of the North Yorkshire Moors Railway – a heritage line – is a special treat for lovers of steam trains and enthusiasts of industrial archaeology. It's also great for those who enjoy strolling through tranquil woods listening to birdsong, all the while accompanied by a beck tumbling gently downstream. The popular route is well-signposted throughout and there are several benches along the way. The walk ends at the lovely village of Goathland.

GO BY TRAIN

OS information

🚶 NZ 828052
Explorer OL27

Distance
3.5 miles/5.5km

Time
2 hours

Start Grosmont
Finish Goathland

Parking
North York Moors National Park pay-and-display car parks: Grosmont, YO22 5QE; Goathland, YO22 5LX

Public toilets
At both car parks; at both North Yorkshire Moors Railway stations

Cafés/pubs
Grosmont; Birch Hall Inn, Beck Hole; Goathland

Did you know?

Goathland is home to a team of Long Sword dancers – the Goathland Plough Stots. Believed to date back to Viking times, this traditional dance takes place in many towns and villages in Yorkshire, where they raise money for local charities. Typically a male pursuit, the Plough Stots caused controversy in 2010 by allowing women to take part. This led to the Stots being barred from the 'Morris Ring', which governs Long Sword and Morris dancing.

Local legend

The woodland around Beck Hole was said to be the home of a giant worm with a mouth like a dragon. It lived on sheep and cattle and had a distinctive smell like wild garlic. Villagers eventually managed to lure it from its lair by tying a maiden, called Kitty, to a tree as bait. The worm was slain by a Danish knight called Roland who then rode off with Kitty.

Terrain	
Mostly surfaced paths	

Hilliness Short climb from ① to ②; then almost flat; steady climb from ⑤ to ⑦; a final, short drop to Goathland Station

Footwear Year round

Public transport

National rail services to Grosmont: nationalrail.co.uk; Grosmont and Goathland linked by North Yorkshire Moors Railway (daily April to November; weekends only in winter): nymr.co.uk

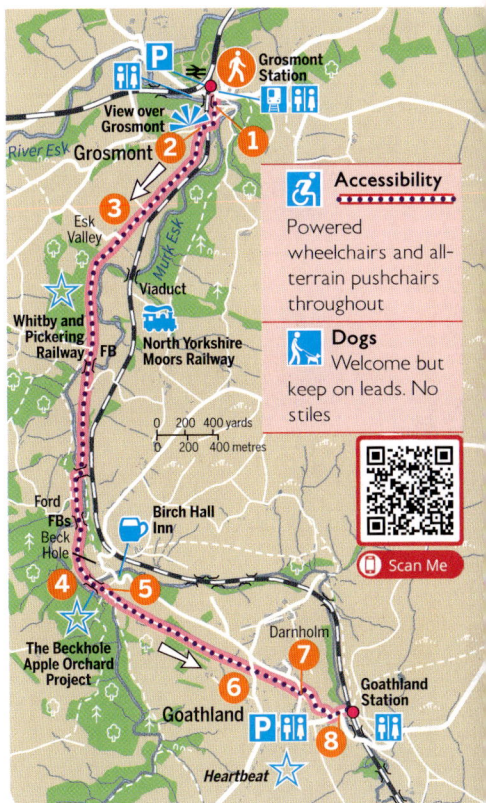

Accessibility

Powered wheelchairs and all-terrain pushchairs throughout

Dogs

Welcome but keep on leads. No stiles

Scan Me

STORIES BEHIND THE WALK

☆ Whitby and Pickering Railway

The original Whitby and Pickering Railway, which runs alongside today's heritage line in places, was a horse-drawn railway designed by George Stephenson, often regarded as the 'father of the railways'. The

incline from **5** to **7**, which has a 1-in-15 gradient, involved a system of ropes and pulleys. Descending carriages, coupled to water tanks for additional weight and speed, provided the motivating force for uphill traffic.

☆ The Beckhole Apple Orchard Project

Arriving by train, Victorian tourists used to stop off at Beck Hole to visit the local tea gardens, set in and around apple orchards. In 2009, members of the Beckhole Woodland and Heritage Foundation decided to restore one of these orchards. The group planted 20 varieties of apple trees, each of which was adopted by a pupil at Goathland Primary School.

North Yorkshire Moors Railway
🚂 Path fork
View over Grosmont
🚶 **1** Wooden bench
Suspension bridge
🚻 Grosmont Station
2
½ mile
Esk Valley
3
☆
W h i t b y   a n d

▶ From the end of Grosmont's Pickering-bound platform, cross the road.
▶ Take the path towards the church, with the railway on your right.
▶ After the suspension bridge, reach a path fork.

1 ▶ Bear **left** at the fork.
▶ Go through the gate at the top of the slope and turn **right**, soon going through another gate.
▶ Continue to a wooden bench with lovely views over Grosmont.

☆ *Heartbeat*

Filmed in and around Goathland, the TV series *Heartbeat* was first broadcast by the BBC in 1992. Based on the 'Constable' novels by Nicholas Rhea and set in the 1960s, the popular drama ran for 18 years and attracted regular audiences of more than 10 million viewers. Goathland became the fictional village of Aidensfield, with series regulars such as Nick Berry sharing the village with A-listers, including Daniel Craig and Benedict Cumberbatch.

🚂 North Yorkshire Moors Railway (NYMR)

Running between Pickering and Grosmont – continuing to Whitby as a scheduled, mainline service – the North Yorkshire Moors Railway is the busiest heritage line in England. The original line opened in 1836, was reengineered and rerouted for steam locomotives in the mid-19th century, and was finally closed in 1965 as part of the cuts recommended by the Beeching Report. The NYMR Preservation Society – formed in 1967 – eventually purchased the line, reopening it to tourists and rail enthusiasts in 1973 (nymr.co.uk).

1 mile

1½ miles

Pickering Railway

2 ➤ At the bench, turn **left** through a gate.
➤ Beyond the next gate, the path swings **right**, and then runs arrow-straight for nearly ½ mile to a crossways in the hamlet of Esk Valley.

3 ➤ Reaching Esk Valley, continue **straight on**.
➤ After almost 1 mile, before the riverside path descends steps, turn **right** to cross the footbridge over the river.
➤ Continue for nearly ½ mile to approach the hamlet of Beck Hole.

NATURE NOTES

Much of this walk follows the route of the Murk Esk, one of the main tributaries of the River Esk. Otters, most likely seen just after dawn, frequent its banks, while dippers can be spotted flying along its channel, flitting from one rock to another. A flash of electric blue could be a kingfisher, but watch too for the blue of the emperor dragonfly and the common blue damselfly. Sea trout spawn upstream and downstream of Beck Hole, while salmon spawn downstream.

As well as passing through broadleaved woodland – home to oak, sycamore, hazel, wild cherry, beech, holly, ash and elder – the paths on this route are often lined with bramble, crab apple, blackthorn, hawthorn, greater burdock, tufted vetch (also known as cow vetch and bird vetch), meadow buttercup, bluebells, wild garlic and dog's mercury.

Tufted vetch

Footbridge over the Murk Esk

Path fork; Beck Hole (left)

Junction with a wider track

2 miles

2½ miles

Birch Hall Inn

The Beckhole Apple Orchard Project

4 ▶ Nearing Beck Hole, bear **right** as the path ahead splits, soon crossing another wooden bridge.
▶ Carry on for 125 yards to a junction with a wider track.

5 ▶ At the junction, turn **right**.
▶ Shortly, after some cottages, the path begins a long, steady climb to Goathland, reaching a road in just over ½ mile.

Above: sloes, the fruit of blackthorn
Below: dipper

Top: otter
Bottom: common blue damselfly

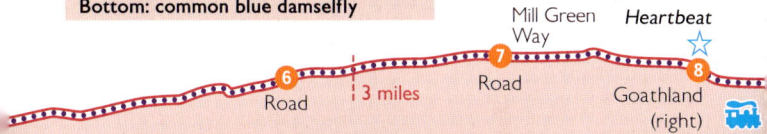

Mill Green Way

Heartbeat

☆

6
Road

3 miles

7
Road

8
Goathland
(right)

🚂

🚻

Goathland
Station

6 ► Go **straight across** the road and through the gate opposite, continuing uphill on a clear path to the next road.

7 ► Turn **left** along the road for 35 yards, and then turn **right** along Mill Green Way.
► At a T-junction, turn **left** along the pavement for 40 yards to the first drop.

8 ► At the bottom of the first drop, go **right** and immediately **left** to descend to Goathland Station.

Opposite (clockwise): Blacksmiths Arms, Lastingham; Whitby Brewery
This page (clockwise): Gliders' Nest Café, Yorkshire Gliding Club; Queen Catherine Hotel, Osmotherley; Park Life Café, North York Moors National Park Centre, Sutton Bank; Daffy Caffy, High Mill, Farndale; ice-cream van, Whitby

WHITBY ABBEY

The atmospheric Gothic ruins of Whitby Abbey are perched on a headland high above the historic harbour town – an impressive spot made all the more dramatic by its links with Bram Stoker's *Dracula*. This walk visits the abbey and neighbouring St Mary's Church before making its way along an easy-going clifftop path in the company of seabirds. Turning away from those magnificent North Sea views, the inland return to the abbey is via field paths and quiet lanes.

OS information

NZ 904110
Explorer OL27

Distance
3.1 miles/4.9km

Time
1¾ hours

Start/Finish
Whitby Abbey

Parking YO22 4EH
Whitby Abbey
Headland pay-and-display car park,
Abbey Lane

Public toilets
In the car park

Cafés/pubs
Whitby Brewery;
seasonal ice-cream van outside abbey;
Whitby

Terrain
Pavements; surfaced clifftop path; lanes; field paths; steps before ③

Hilliness
Gently undulating

Footwear
Winter/ Spring/ Autumn 🥾
Summer 🥾

Public transport
None to the abbey/ 🚶; national rail and regional bus services to Whitby town centre, ½ mile from ① – up the 199 Steps: nationalrail.co.uk; northyorkstravel.info

Did you know? Whitby gives its name to a form
of organic gemstone, Whitby jet, which is hard
enough to be used for jewellery making. By the
middle of the 19th century, there were around 50
workshops producing jet ornaments for tourists
eager to remember their visit to the town.
After the death of Prince Albert in 1861, Queen
Victoria adorned her mourning clothes with
Whitby jet, further increasing its popularity.

Local legend Bram Stoker visited the town in
1890 and was inspired by the abbey and St Mary's
churchyard to develop his Gothic masterpiece. It
was supposedly a visit to the local library that
led to the discovery of the name 'Dracula' in a
book written by the British Consul in Bucharest.
Dracula's ship, the *Demeter*, may have taken its
name from the Russian ship *Dmitry*, which ran
aground below East Cliff in 1885.

STORIES BEHIND THE WALK

Whitby Abbey There's evidence of occupation on the East Cliff headland dating back to the Bronze Age. The first monastery was founded in the mid-seventh century, but this was later abandoned. A new monastic community was founded in the 11th century and later developed into a Benedictine monastery. The current Gothic ruins date to a remodelling of the buildings between the 13th and 16th centuries.

England Coast Path When complete, the King Charles III England Coast Path will follow the coastline of England for more than 2,500 miles, making it the longest managed coastal path in the world. A 2009 Act of Parliament instructed Natural England to set up a continuous path around the English coast and the first official section was opened in Dorset in 2012. Originally called the England Coast Path, it was rebranded to mark the King's coronation in 2023.

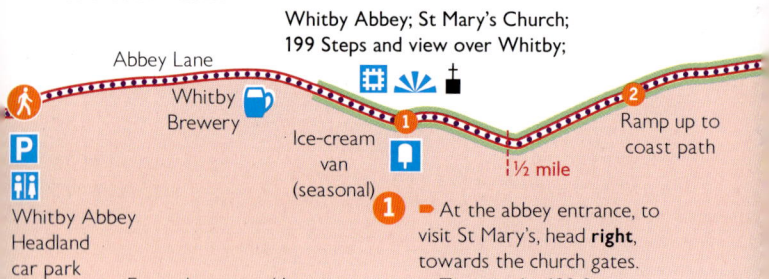

Whitby Abbey; St Mary's Church; 199 Steps and view over Whitby;

Abbey Lane

Whitby Brewery

Ice-cream van (seasonal)

Ramp up to coast path

½ mile

Whitby Abbey Headland car park

1 ➤ From the car park's main entrance, turn **left** along the pavement beside the narrow lane.
➤ The pavement later swings **left** and then **right**, ending near the abbey entrance.

1 ➤ At the abbey entrance, to visit St Mary's, head **right**, towards the church gates.
➤ To view the 199 Steps, take the **left-hand** path after the gates.
➤ Having explored the church, retrace your route 175 yards up Abbey Lane to the coast path turning.

☀ View of Whitby

From the top of the 199 Steps and from St Mary's churchyard, visitors have an extensive view over much of the old town, with its buildings clustered around the mouth of the River Esk. The name Whitby comes from the Norse for white settlement. The town developed from a fishing port to become an important centre for ship building and whaling before turning into a popular tourist destination in Victorian times.

✝ St Mary's Church

St Mary's sits at the top of the 199 Steps, high above the town. Built in the early 12th century, the church was significantly expanded in 1818. The surrounding graveyard featured in a scene from Bram Stoker's book *Dracula* and, more recently, in the video for Simply Red's *Holding Back the Years*. In 2012, heavy rain washed away part of the graveyard and human bones were found in a street at the base of the headland.

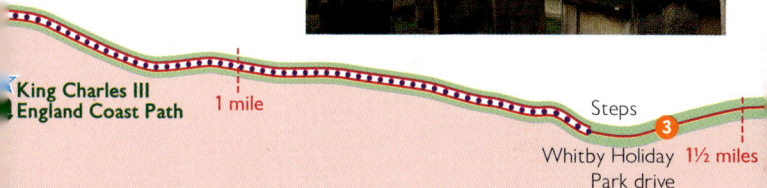

◀ King Charles III England Coast Path

1 mile

Steps

③

Whitby Holiday Park drive

1½ miles

② ▶ Signposted England Coast Path, go **left** up the ramp.
▶ The coast path bends **right**, passes to the left of houses and then heads along the clifftop for just over ¾ mile to reach an asphalt drive in a holiday park.

③ ▶ Go **left** along the drive for 200 yards, passing to the right of the park's reception buildings, to reach a split in the driveway on the far side of the site.

NATURE NOTES

The cliffs along Yorkshire's North Sea coast are popular nesting spots for a variety of seabirds. The kittiwake, for example, nests on ledges on the cliff face from February until August. It then spends the winter out at sea. Other seabirds that can be spotted from these elevated cliffs include fulmars and herring gulls. Gannets are also present. Known for their high-speed dives – they circle high above the sea, fold back their wings and then plunge headfirst into the water, reaching speeds of up to 60mph. Inland, bird species in the fields and along the lanes include wood pigeons, rooks, goldfinches, robins, wrens and blue tits, among others.

In the height of summer, watch for spear thistles fringing the clifftop path as well as common mallow with its pretty pink flowers. The latter has medicinal properties and is used to treat dry coughs, sore throats, mild inflammation and stomach complaints.

Gannet

King Charles III England Coast Path

4 Driveway split

5 T-junction with a road

2 miles

4 ➤ Bear **left** as the driveway splits.
➤ In a further 50 yards, keep **straight** ahead as the coast path peels off to the left, soon rejoining the asphalt lane.
➤ Keep **ahead** for ⅓ mile to a T-junction with a road.

5 ➤ Turn **right** at the T-junction and walk along the road for 400 yards.
➤ Take the signposted public footpath through the hedgerow gap on the **left**. Keep to the field's left-hand edge and leave via a small gate onto a lane.

Top left: fulmar
Above: common mallow
Bottom left: kittiwake
Below: spear thistle

Abbey Lane

8

3 miles

T-junction

7

2½ miles

6

Small gate
onto a lane

Whitby Abbey
Headland
car park

7 ▸ Turn **left** at the
junction and then,
after a few strides,
go **right** at the next
T-junction.
▸ Head along the
pavement of Green
Lane for ¼ mile to
another T-junction.

8 ▸ At the junction
with Abbey Lane,
turn **left** – towards
the abbey and youth
hostel.
▸ The car park
entrance is on the
left in 75 yards.

6 ▸ Keep **straight
ahead** along the lane.
This later bends
right and ends at a
T-junction.

WALK 9

CATCH A BUS

ROBIN HOOD'S BAY

This is a walk of two halves. Starting from the popular seaside village of Robin Hood's Bay, it first heads out along a disused railway – a shared-use path known as the Cinder Track – with good views out to sea. For the second half, it drops onto the England Coast Path for a walk along high, windswept cliffs. Although the route doesn't visit the lower, older part of the village, there are some impressive glimpses of it on this return leg.

OS information

🏃 NZ 949054
Explorer OL27

Distance
3.1 miles/5km

Time
1¾ hours

Start/Finish
Robin Hood's Bay

Parking YO22 4RA
Station Road car park (pay-and-display)

Public toilets
In the car park

Cafés/pubs
Robin Hood's Bay

Terrain
Pavements; surfaced, shared-use old railway path; good clifftop path; some steps

Hilliness
Almost flat until ❸; short, steep drop to ❹; gently undulating coast path

Footwear
Winter/ Spring/ Autumn 👢
Summer 👟

🚌 **Public transport**
Bus service X93, Middlesbrough to Scarborough, stops in Thorpe Lane, 350 yards south of 🚶: arrivabus.co.uk

Did you know? Next to the slipway in Robin Hood's Bay is a cast-iron collection box in the shape of a cod fish standing on its tail with its open mouth forming a money slot. It was installed in 1887 to collect funds for the Royal National Lifeboat Institution and is now the charity's oldest collection box. Nearly four feet tall and weighing about 17 stones, it is one of Britain's smallest Grade II-listed structures.

Local legend A mile south of Robin Hood's Bay lies the small cove of Boggle Hole. Boggles were believed to be a type of hobgoblin that lived in holes and caves. The one on this section of the coast was believed to have healing powers. Mothers would often bring their children down from the villages to be cured of a range of illnesses.

Accessibility

Suitable for robust, powered wheelchairs and all-terrain pushchairs 🚶 to ③; also, from the Rocket Post to the car park, enabling a there-and-back visit to the clifftop viewpoint

Dogs Welcome but keep on leads. No stiles

Far Jetticks

Clock Case Nab

④

③ Waterfalls

⑤ Rain Dale

Craze Naze

Homerell Hole

Cow & Calf

King Charles III England Coast Path

The Cinder Track

Coast to Coast Path

Castle Chamber

Bulmer Steel

Bulmer Steel Hole

Ness Point or North Cheek

Waterfall

Ness Ruck

Bay Ness

Smails Moor Farm

Green Hills

B1447

Hook's House

Lingers Beck

② ⑦

⑥

Dungeon Hole

Rocket Post

View to Robin Hood's Bay

①

PO

Robin Hood's Bay

Ground Wyke

West Scar

0   200   400   600 yards
0   200   400   600 metres

Scan Me

STORIES BEHIND THE WALK

🔷 **Robin Hood's Bay** Visible from a viewpoint near the Rocket Post, Robin Hood's Bay was one of the busiest smuggling ports in the country. In the 18th century, the village's isolated position made it an ideal place to dodge the excise men. Secret passageways and tunnels below and between the houses made it easy for villagers to transfer illegal goods. If the excise men ventured onto the narrow streets, they were often doused with boiling water from the windows above.

☆ **Coast to Coast Path** The walk's second half follows the closing stages of the 190-mile Coast to Coast Path – one of Britain's most popular long-distance trails. It starts in St Bees in Cumbria and ends at Robin Hood's Bay, where it's traditional for hikers to dip their toes in the North Sea. Alfred Wainwright published his guide to the Coast to Coast in 1973. He suggested walkers should find their own routes across the country, but it becomes a National Trail in 2025.

Road junction | Turning onto the Cinder Track | ☆ **The Cinder Track**

½ mile

Mount Pleasant North
🅿️
🚻
Station Road car park

➤ Standing at the car park entrance, with your back to the toilet block, turn **left** along Station Road and walk to the T-junction **ahead**.

1 ➤ At the T-junction, go **left** and immediately **right** – along Mount Pleasant North (MPN).
➤ Towards the end, MPN narrows and, ignoring the turning right, climbs left – follow Cinder Track sign – to the second turning right at the top.

Rocket Post Located in the clifftop field to the north of Robin Hood's Bay, the Rocket Post is a replica of a piece of life-saving equipment. If a ship got into difficulties on the coast below, a rocket would be fired carrying a rope which could be secured to the ship's mast. A breeches buoy (canvas shorts attached to a cork lifebuoy) would then be sent along the line to recover stranded sailors.

☆ **The Cinder Track** The first half of the walk follows part of the Cinder Track, a 21-mile walking and cycling path along the trackbed of the Scarborough to Whitby railway – closed in 1965 as part of the Beeching cuts. In 2018, plans to improve and upgrade the path were approved, including new drainage and resurfacing. The Cinder Path is part of the North Sea Cycle Route, which follows the coastline of much of northern Europe including Norway, Denmark and Holland.

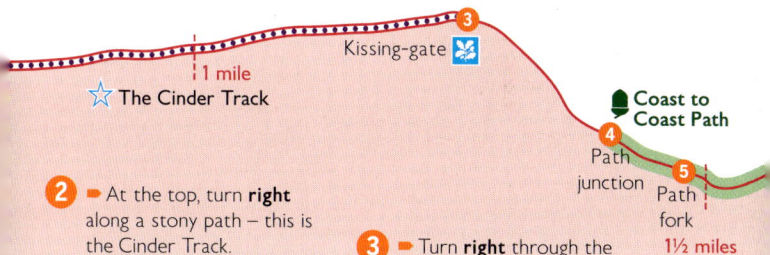

1 mile

☆ **The Cinder Track**

Kissing-gate

Coast to Coast Path

④ Path junction

⑤ Path fork

1½ miles

② ➤ At the top, turn **right** along a stony path – this is the Cinder Track.
➤ Follow the old railway trackbed for just over 1 mile to a kissing-gate and National Trust Bay Ness sign (right).

③ ➤ Turn **right** through the kissing-gate and follow a grassy path that heads steeply downhill to a path junction at the bottom.

NATURE NOTES

Walk the Cinder Track in spring and summer and you're likely to see butterflies such as small copper, peacock and common blue. The caterpillars of these butterflies feed on the plants that grow alongside the track: common sorrel and dock (small copper), stinging nettle (peacock) and bird's-foot trefoil (common blue). Other trees, shrubs and wildflowers lining the route include bramble, red campion, elder, ragwort, hawthorn, great willowherb, red clover and hemp agrimony, also known as 'raspberries and cream' because of its clusters of pink flowers.

If you've got binoculars, look out to sea from the cliffs for a chance of spotting bottlenose dolphins and even minke whales. Easier to spot are the badger face sheep that sometimes graze the Rocket Post field. Named for the distinctive black stripes above their eyes, this hardy upland breed is more commonly associated with the Welsh mountains.

Top: hemp agrimony
Middle: elderflower
Bottom: minke whale

🌲 Coast to
Coast Path

2 miles

4 ➤ At the junction, turn **right** along a stony path. This is the King Charles III England Coast Path.
➤ Carry on (sea on the left) to a fork in about 125 yards.

5 ➤ At the fork, keep **left** to walk beside the cliff edge.
➤ The grassy path undulates along the clifftop for 1¼ miles, becoming surfaced by the Rocket Post.
➤ About 250 yards beyond the Rocket Post, the path bends sharp **right** to a gate.

Badger face sheep

Top: small copper
Middle: common blue (male)
Bottom: common blue (female)

View to Robin Hood's Bay 2½ miles

Rocket Post (right)

Path junction with the Cinder Track

Mount Pleasant North

⑥ Path bends sharp right to a gate

⑦ 3 miles

Station Road car park

⑥ ➤ Pass through the gate and a second gate, directly **ahead**.
➤ After the second gate, the path bends **left** and shortly drops to regain the Cinder Track.

⑦ ➤ On meeting the Cinder Track, bear **left** and retrace your steps to the Station Road car park.

BROXA FOREST

Broxa Forest is located in the south-eastern corner of the National Park, on the edge of an area dominated by forestry. Don't expect dark, oppressive plantations though; there is mixed woodland here, as well as open areas with great views over Harwood Dale. This easy walk combines broad tracks, sometimes shared with mountain bikers, and quieter trails, where there's a good chance that your only company will be woodland birds and, if you're lucky, an occasional deer.

OS information

SE 965943
Explorer OL27

Distance
3.4 miles/5.5km

Time
2 hours

Start/Finish
Broxa Forest

Parking YO13 0JT
Forestry England
Broxa Forest car
park at Reasty Hill
Top (charge)

Public toilets
None

Cafés/pubs
None

Terrain
Mostly on broad,
surfaced forest
tracks; narrow,
compacted-earth
trail

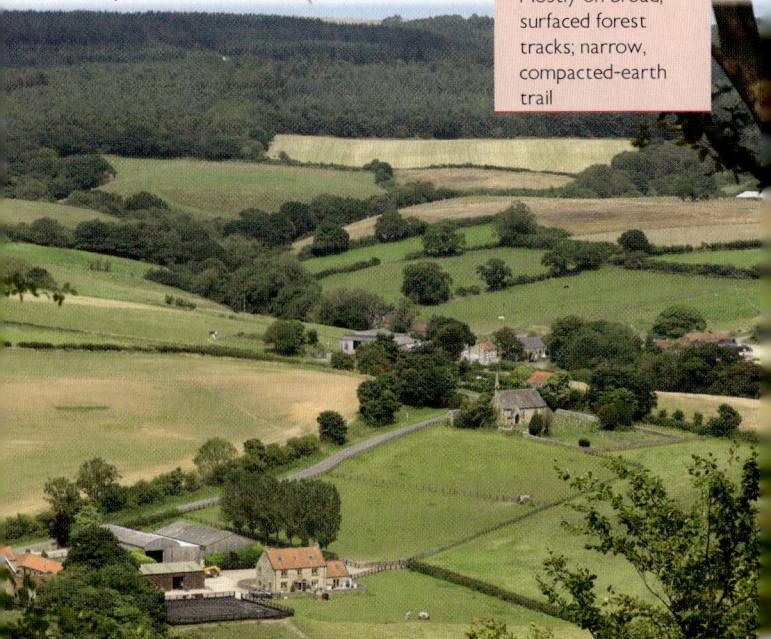

Hilliness
Fairly flat; one gentle descent **2** to **5**, and a gentle rise to **7**

Footwear
Winter/ Spring/ Autumn 👢
Summer 👟

🚌 Public transport
None

♿ Accessibility
Powered, robust wheelchairs 🚶 to **4**, with return by outward route; suitable for all-terrain pushchairs throughout

🐕 Dogs
Welcome.
No stiles

Did you know? The ruined church of St Margaret's in Harwood Dale, to the north of Broxa Forest, was built in memory of Lady Margaret Hoby, author of the earliest surviving diary written in English by a woman. The diary covers the period from 1599 to 1605 and details her religious pursuits and prayers as well as occasional glimpses of her daily life.

Local legend In 1957, three men found a strange metal object, believed to be a UFO, on Broxa Forest's Silpho Moor. The base of it was covered in hieroglyphs – which some thought was Cyrillic script – and a metal 'book' was found inside, also containing hieroglyphs. The object disappeared soon after its discovery but reappeared in London's Science Museum in 2018. Metallurgists determined that the object had never been to outer space and its 'discovery' was a hoax.

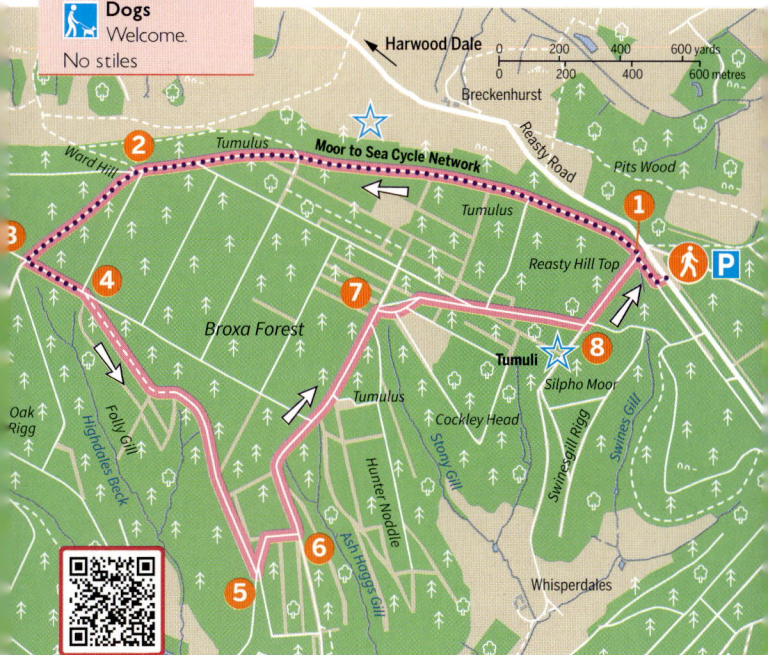

Scan Me

Walk 10 Broxa Forest **75**

STORIES BEHIND THE WALK

☆ **Forestry England history** Broxa Forest is managed by Forestry England, formerly known as the Forestry Commission. The commission was set up after World War I, during which woodland resources had been severely depleted, particularly by trench warfare. This meant there was a massive need to rebuild and maintain a strategic timber reserve. The Forestry Act came into force in 1919, giving the new commission a lot of freedom to acquire and plant land.

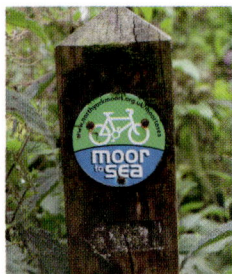

☆ **Moor to Sea Cycle Network** Part of the walk coincides with the Moor to Sea Cycle Network, a long-distance cycle trail connecting Scarborough, Whitby, Dalby Forest, Pickering and Great Ayton. The network covers around 150 miles of tracks, bridleways and quiet roads in the north and east of the North York Moors National Park. The whole route can be completed in a long day, but loops of varying length mean it can be explored in shorter stages.

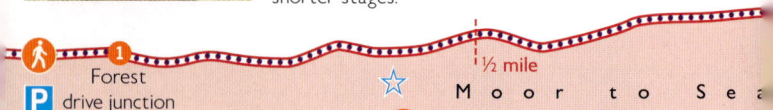

1 Forest drive junction

Forestry England Broxa Forest car park (Reasty Hill Top)

➡ From the car park's vehicle entrance, cross the road **diagonally left** and head into the forest.

➡ The broad forest drive quickly bends **right** and soon leads to a junction.

½ mile

M o o r   t o   S e a

1 ➡ Bear **right** at the junction.

➡ Keeping to the main track, ignore another track going sharp left after ¾ mile, and stay **forward** for a further ¼ mile to reach a fork (with a barrier across the right-hand track).

There are a number of different types of barrows dotted throughout Broxa Forest. Many of these are Bronze Age round barrows, common in England, but others are rarer Iron Age 'square' barrows. These single-burial mounds are most common in North and East Yorkshire and are surrounded by square ditches, sometimes with a bank. Many have been found to contain high-status grave goods. This walk passes close to at least two of these, near **8**.

☆ **Forestry England today** The size of the forests to the south and west of Broxa becomes apparent from views gained between **4** and **5**. Although they were originally established for strategic reasons after World War I, their management today combines timber production, leisure and ecological considerations. Nearby Cropton Forest, for example, is the site of a trial in natural flood management using reintroduced beavers, while Dalby Forest is criss-crossed by mountain biking trails designed for all abilities (forestryengland.uk/broxa-forest).

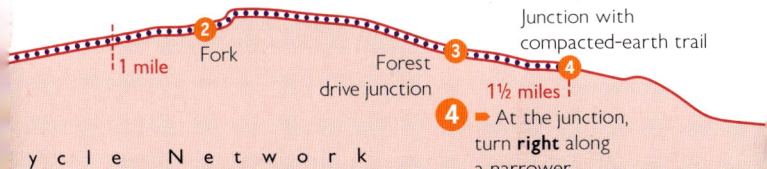

Junction with compacted-earth trail

2 Fork

1 mile

3 Forest drive junction

1½ miles

4

y c l e   N e t w o r k

2 ► Ignore the option with the barrier, and follow the main track to the **left** for almost ⅓ mile to the next main forest drive junction.

3 ► At the junction, turn **left** onto a track with a barrier across it.
► Keep **forward** to the next junction in just over 200 yards.

4 ► At the junction, turn **right** along a narrower, compacted-earth trail.
► Follow this for about ⅔ mile, emerging from a darker part of the forest to reach a T-junction of trails.

Walk 10 Broxa Forest   77

NATURE NOTES

Although managed by Forestry England, Broxa Forest is more than just a conifer plantation. The walk passes through mixed woodland that also contains oak, birch, sycamore, beech, rowan and sweet chestnut. Watch for well-camouflaged treecreepers spiralling up the tree trunks, using their long, down-curved bills to search the bark for insects.

When autumn leaves cover the woodland floor and moisture begins to build up, look for fungi such as puffballs, shaggy inkcaps, stinkhorn and orange peel fungus. The distinctive red caps of fly agaric are most common beneath stands of Scots pine.

Ground flora includes common spotted orchid, yellow loosestrife, meadow buttercup, selfheal and, in more open glades, bell heather, bilberry and wild marjoram. Bearing pink flowers in summer, the marjoram is the same species as the Mediterranean herb oregano but it has a slightly different scent in Britain's cooler climate.

Rowan berries

Trail
T-junction

Wide Forest
track

2 miles

5

6

2½ mile

5 ➤ Turn **left** at the T-junction.
➤ The path then bends sharp **right**, soon reaching a wide forest track.

6 ➤ Turn **left** along the wide forest track, following it for ½ mile to a junction where a wide grassier track continues straight on.

Top left: yellow loosestrife
Top right: treecreeper
Above: wild majoram
Left: orange peel fungus

3 miles

☆ Tumuli

T-junction just beyond a barrier

S i l p h o   M o o r

🅿 Forestry England Broxa Forest car park (Reasty Hill Top)

7 ➡ Ignore the grassier straight on option and follow the clearer track to the **right**.
➡ This quickly swings **left** and then bears **right** before running straight for ⅓ mile to a track T-junction, just beyond a barrier.

8 ➡ Go **left** at the T-junction of tracks.
➡ In another 275 yards, rejoin your outward route at the junction **1**.
➡ Turn **right** and retrace your steps to the car park.

Publishing information

ISBN 978 0 319092 88 0
1st edition published by Ordnance Survey 2025.

ordnancesurvey.co.uk

While every care has been taken to ensure the accuracy of the route directions, the publishers cannot accept responsibility for errors or omissions, or for changes in details given. The countryside is not static: hedges and fences can be removed, stiles can be replaced by gates, field boundaries can alter, footpaths can be rerouted and changes in ownership can result in the closure or diversion of some concessionary paths. Also, paths that are easy and pleasant for walking in fine conditions may become slippery, muddy and difficult in wet weather.

If you find an inaccuracy in either the text or maps, please contact Ordnance Survey at os.uk/contact.

Milestone Publishing credits

Author: Vivienne Crow

Series editor: Kevin Freeborn

Maps: Cosmographics

Design and Production: Patrick Dawson, Milestone Publishing

Printed in India by Replika Press Pvt. Ltd

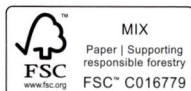

MIX
Paper | Supporting responsible forestry
FSC™ C016779

Photography credits

Front cover: Dave Porter/Alamy Stock Photo
Back cover cornfield/Shutterstock.com.

All photographs supplied by the author ©Vivienne Crow except page 6 Tom and Essi Troughton (Ordnance Survey).

The following images were supplied by

Alamy Stock Photo: page 47 Les Gibbon.

Public Domain, via Wikimedia Commons: page 59 Davepape.

CC BY-SA 4.0, via Wikimedia Commons: page 53 Dunpharlain; 39 Michel Langeveld; 53 xpda; 59 Олексій Карпенко; 67, 73 Charles J. Sharp; 72 Cephas.

CC BY-SA 3.0, via Wikimedia Commons: page 46 Charles J. Sharp; 59 L. B. Tettenborn.

CC BY-SA 2.0, via Wikimedia Commons: page 53 Amanda Slater; 33 David Hawgood; 33 Mick Lobb; 40 Patrick Roper; 27 Paul Buckingham; 32 Peter Pearson; 38 Philip Halling; 79 RHL Images; 38 Richard Dorrell; 19 Robn Webster; 33 Russel Wills; 18 Walter Baxter; 46 Alexis Lours; 66 inkasaur; 79 Holger Krisp.